P. Mangani (Ed.)

Model Theory and Applications

Lectures given at a Summer School of the
Centro Internazionale Matematico Estivo (C.I.M.E.),
held in Bressanone (Bolzano), Italy,
June 20-28, 1975

Springer

FONDAZIONE CIME ROBERTO CONTI

C.I.M.E. Foundation
c/o Dipartimento di Matematica "U. Dini"
Viale Morgagni n. 67/a
50134 Firenze
Italy
cime@math.unifi.it

ISBN 978-3-642-11119-8 e-ISBN: 978-3-642-11121-1
DOI:10.1007/978-3-642-11121-1
Springer Heidelberg Dordrecht London New York

Reprint of the 1st ed. C.I.M.E., Ed. Cremonese, Roma 1975
With kind permission of C.I.M.E.

Printed on acid-free paper

Springer.com

To the Memory of Abraham Robinson

CENTRO INTERNAZIONALE MATEMATICO ESTIVO

(C.I.M.E.)

2° Ciclo - Bressanone dal 20 al 28 giugno 1975

MODEL THEORY AND APPLICATIONS

Coordinatore: Prof. P. MANGANI

Centro Internazionale Matematico Estivo

"Model Theory and Applications"

(Second 1975 C. I. M. E. Session)

Lecture Notes for Course (a)

Theories of Algebraic Type

Gerald E. Sacks

Bressanone (Bolzano), Italy

1. Fundamentals

Buon giorno. This is the first of eight lectures on the model theoretic notion of theory of algebraic type. Some examples of the notion are the theories of algebraically closed fields of characteristic p ($p \geq 0$), real closed fields and differentially closed fields of characteristic 0. The last example is the most important for two reasons. First, it is the only one known whose complexity matches that of the general case. Second, several results about differential fields, results which hold for all theories of algebraic type, were first proved by model theoretic means.

The key definition is quite compact, but five lectures will be needed to unpack it. A theory T is said to be of <u>algebraic type</u> if T is complete, T is the model completion of a universal theory, and T is quasi-totally transcendental. In the brief time left before the onset of formalities, let me indicate why the theory of algebraically closed fields of characteristic 0 (ACF_0) is of algebraic type. The completeness of ACF_0 means that the same first order sentences are true in all algebraically closed fields of characteristic 0. Thus a first order sentence in the language of fields is true of the complex numbers if and only if it is true of the algebraic numbers.

ACF_0 is the model completion of TF_0, the theory of fields of characteristic 0. To say TF_0 is a universal theory is equivalent to saying a subset of a field of characteristic 0 closed under $+, \cdot$, etc. is a field of characteristic 0. To see the meaning of model completion, let $\mathcal{A}$ be any field of characteristic 0 and let F be any first order sentence in the language of fields with parameters in $\mathcal{A}$. (For example, F might say that some finite set of polynomials in several variables with coefficients in $\mathcal{A}$ has a common zero.) To claim that ACF_0 is the model completion of TF_0 amounts to claiming F is true in all or in none of the algebraically closed extensions of $\mathcal{A}$.

The property of quasi-total transcendality is too complex to elucidate in a lecture on fundamentals. For the moment think of it as a density condition on simply generated extensions of structures weakly exemplified by the density of the rationals in the reals. If T is quasi-totally transcendental, then each substructure $\mathcal{A}$ of a model of T has a prime model extension, and all prime model extensions of $\mathcal{A}$ are isomorphic over $\mathcal{A}$. In the case of ACF_0, this means each field $\mathcal{A}$ of characteristic 0 has a unique prime algebraically closed extension, namely the algebraic closure of $\mathcal{A}$.

And now the fundamentals of model theory. A <u>similarity type</u> τ is a 5-tuple $\langle I, J, K, \theta, \psi \rangle$ such that $\theta : I \to N$ and $\psi : J \to N$, where N is the set of positive integers. A <u>structure</u> $\mathcal{A}$ of type τ consists of:

(i) A nonempty set A called the universe of $\mathcal{A}$.

(ii) A family $\{R_i^{\mathcal{A}} \mid i \in I\}$ of relations. Each $R_i^{\mathcal{A}}$ is a subset of $A^{\theta(i)}$.

(iii) A family $\{f_j^{\mathcal{A}} \mid j \in J\}$ of functions. Each $f_j^{\mathcal{A}}$ maps $A^{\psi(j)}$ into A.

(iv) A subset $\{c_k^{\mathcal{A}} \mid k \in K\}$ of A called the set of distinguished elements of A.

One often writes

$$\mathcal{A} = \langle A, R_i^{\mathcal{A}}, f_j^{\mathcal{A}}, c_k^{\mathcal{A}} \rangle_{i \in I,\, j \in J,\, k \in K} .$$

The cardinality of $\mathcal{A}$ is by definition the cardinality of A. Structures will be denoted by $\mathcal{A}, \mathcal{B}, \mathcal{C}, \ldots$, and their universes by $A, B, C, \ldots$.

Consider the structure

$$\mathcal{A} = \langle A, +, \cdot, -, {}^{-1}, 0, 1 \rangle ,$$

where $+$ and $\cdot$ are 2-place functions on A, $-$ and $^{-1}$ are 1-place functions on A, and 0 and 1 are distinguished elements of A. The concept of field can be formulated so that every field has the same similarity type as $\mathcal{A}$, but $\mathcal{A}$ need not be a field since the relations, functions and distinguished elements of $\mathcal{A}$ need not satisfy the axioms for fields.

A monomorphism $m : \mathcal{A} \to \mathcal{B}$ is a one-one map $m : A \to B$ such that:

(i) $R_i^{\mathcal{A}}(a_1, \ldots, a_n)$ iff $R_i^{\mathcal{B}}(ma_1, \ldots, ma_n)$ ($i \in I$ and $n = \theta(i)$).

(ii) $mf_j^{\mathcal{A}}(a_1, \ldots, a_n) = f_j^{\mathcal{B}}(ma_1, \ldots, ma_n)$ ($j \in J$ and $n = \psi(i)$).

(iii) $mc_k^{\mathcal{A}} = c_k^{\mathcal{B}}$ ($k \in K$).

(It is assumed that $\mathcal{A}$ and $\mathcal{B}$ are both of type τ.)

$\mathcal{A}$ is a <u>substructure</u> of $\mathcal{B}$ ($\mathcal{A} \subset \mathcal{B}$) if $A \subset B$ and the inclusion map $i_A : A \subset B$ is a monomorphism. An <u>isomorphism</u> is an onto monomorphism. An isomorphism is indicated by $m : \mathcal{A} \overset{\approx}{\to} \mathcal{B}$ or by $\mathcal{A} \approx \mathcal{B}$.

Each similarity type τ gives rise to a first order language $\mathcal{L}_\tau$ whose sentences are interpretable in structures of type τ. The primitive symbols of $\mathcal{L}_\tau$ are:

(i) first order variables $x, y, z, \ldots$;

(ii) logical connectives $\sim$ (not), & (and), E (there exists), and = (equals);

(iii) a $\theta(i)$-place relation symbol R_i $(i \in I)$;

(iv) a $\psi(j)$-place function symbol f_j $(j \in J)$;

(v) an individual constant $\underline{c}_k$ $(k \in K)$.

The terms of $\mathcal{L}_\tau$ are generated by two rules: all variables and individual constants are terms; if f_j is an n-place function symbol and $t_1, \ldots, t_n$ are terms, then $f_j(t_1, \ldots, t_n)$ is a term.

The atomic formulas are: equations such as $t_1 = t_2$, where t_1 and t_2 are terms; and $R_i(t_1, \ldots, t_n)$, where R_i is an n-place relation symbol and $t_1, \ldots, t_n$ are terms.

The formulas are generated from the atomic formulas as follows: if F and G are formulas, then $\sim F$, F & G and (Ex)F are formulas, where x is any variable.

$\vee$ (or), $\rightarrow$ (implies), $\leftrightarrow$ (if and only if), and (x) (for all x) are abbreviations: $F \vee G$ for $\sim(\sim F \,\&\, \sim G)$, $F \rightarrow G$ for $(\sim F) \vee G$, $F \leftrightarrow G$ for $(F \rightarrow G) \,\&\, (G \rightarrow F)$, and (x)F for $\sim(Ex)\sim F$.

The predicate, x is a free variable of the formula F, is defined by recursion on the number of steps needed to generate F: if F is atomic and x occurs in F, then x is a free variable of F; if x is a free variable of F, then x is a free variable of $\sim$F, of F & G and of G & F; if x is a free variable of F and y is a variable distinct from x, then x is a free variable of (Ey)F.

The only way to kill a free variable x of F is to prefix F with (Ex). A useful convention is: all the free variables of G(x, y, z) lie among x, y, z.

A sentence is a formula with no free variables.

Each sentence of $\mathcal{L}_\tau$ has a definite truth value in each structure $\mathfrak{A}$ of type τ. As an aid in defining truth, consider the language $\mathcal{L}_{\tau A}$ obtained by adding a new individual constant $\underline{a}$ for each $a \in A$ to the language $\mathcal{L}_\tau$. The formulas of $\mathcal{L}_{\tau A}$ are merely the formulas of $\mathcal{L}_\tau$ with some of the free variables replaced by individual constants naming elements of A. Each constant term (no variables) t of $\mathcal{L}_{\tau A}$ names some element σt of A as follows:

(i) $\sigma\underline{a} = a$ and $\sigma\underline{c}_k = c_k^{\mathfrak{A}}$.

(ii) $\sigma f_j(t_1, \ldots, t_n) = f_j^{\mathfrak{A}}(\sigma t_1, \ldots, \sigma t_n)$.

Let H be a sentence of $\mathcal{L}_{\tau A}$. The relation $\mathfrak{A} \models H$ (H is true in $\mathfrak{A}$) is defined by recursion on the number of steps needed to generate H from the atomic formulas of $\mathcal{L}_{\tau A}$:

$\mathfrak{A} \models t_1 = t_2$ iff $\sigma t_1 = \sigma t_2$.

$\mathfrak{A} \models R_i(t_1, \ldots, t_n)$ iff $R_i^{\mathfrak{A}}(\sigma t_1, \ldots, \sigma t_n)$.

$\mathfrak{A} \models F \,\&\, G$ iff $\mathfrak{A} \models F$ and $\mathfrak{A} \models G$.

$\mathfrak{A} \models \sim F$ iff it is not the case that $\mathfrak{A} \models F$.

$\mathfrak{A} \models (Ex)F(x)$ iff $\mathfrak{A} \models F(\underline{a})$ for some $a \in A$.

If the sentence H is not true in $\mathfrak{A}$, then it is said to be false.

$<a_1, \ldots, a_n>$ satisfies (or realizes) $F(x_1, \ldots, x_n)$ in $\mathfrak{A}$ if $\mathfrak{A} \models F(\underline{a}_1, \ldots, \underline{a}_n)$.

It is now quite simple to say what a field is. The similarity type of a field is exemplified by the structure $\mathfrak{A}$:

$$<A, +^{\mathfrak{A}}, \cdot^{\mathfrak{A}}, -^{\mathfrak{A}}, (-1)^{\mathfrak{A}}, 0^{\mathfrak{A}}, 1^{\mathfrak{A}}>.$$

The nonlogical primitive symbols of the language associated with the similarity type of fields are: $+, \cdot, -, {}^{-1}$, 0 and 1. The theory of fields (TF) is the following set of sentences:

$$(x)(y)(z)[(x+y)+z = x+(y+z)].$$

$$(x)[x+0 = x].$$

$$(x)[x+(-x) = 0].$$

$$(x)(y)[x+y = y+x].$$

$$(x)(y)(z)[(x \cdot y) \cdot z = x \cdot (y \cdot z)].$$

$$(x)[x \cdot 1 = x].$$

$$(x)[x \neq 0 \rightarrow x \cdot x^{-1} = 1].$$

$$(x)(y)[x \cdot y = y \cdot x].$$

$$(x)(y)(z)[x \cdot (y+z) = (x \cdot y)+(x \cdot z)].$$

$$0 \neq 1.$$

$\mathfrak{A}$ is a field iff it has the similarity type specified above and every sentence of TF is true in $\mathfrak{A}$.

$\mathfrak{A}$ is first order (or elementarily) equivalent to $\mathfrak{B}$ ($\mathfrak{A} \equiv \mathfrak{B}$) means: $\mathfrak{A} \models F$ iff $\mathfrak{B} \models F$ for every sentence F. (It is assumed that

$\mathfrak{A}$ and $\mathfrak{B}$ belong to the same similarity type τ, and that F is a sentence of $\mathcal{L}_\tau$.) In the next lecture it will be seen that any two algebraically closed fields of the same characteristic are first order equivalent. More generally it will be observed that any two models of a theory of algebraic type are first order equivalent.

An <u>elementary monomorphism</u> $m : \mathfrak{A} \xrightarrow{\equiv} \mathfrak{B}$ is a map of A into B such that

$$\mathfrak{A} \models F(\underline{a}_1, \ldots, \underline{a}_n) \text{ iff } \mathfrak{B} \models F(\underline{ma}_1, \ldots, \underline{ma}_n)$$

for every formula $F(x_1, \ldots, x_n)$ and every sequence $a_1, \ldots, a_n \in A$. An elementary monomorphism m is necessarily a monomorphism, since

$$\mathfrak{A} \models \underline{a}_1 = \underline{a}_2 \text{ iff } \mathfrak{B} \models \underline{ma}_1 = \underline{ma}_2 .$$

Note that a map m of A into B is an elementary monomorphism of $\mathfrak{A}$ into $\mathfrak{B}$ iff

$$<\mathfrak{A}, a>_{a \in A} \equiv <\mathfrak{B}, ma>_{a \in A} .$$

(The similarity type of $<\mathfrak{A}, a>_{a \in A}$ is $\mathcal{L}_{\tau A}$.)

Proposition 1. Suppose $f : \mathfrak{A} \to \mathfrak{B}$ and $g : \mathfrak{B} \to \mathfrak{C}$.

(i) If f and g are elementary, then gf is elementary.

(ii) If g and gf are elementary, then f is elementary.

$\mathfrak{A}$ is an <u>elementary substructure</u> of $\mathfrak{B}$ (or $\mathfrak{B}$ is an <u>elementary extension</u> of $\mathfrak{A}$) if $\mathfrak{A}$ is a substructure of $\mathfrak{B}$ and the inclusion map $i_A : A \subset B$ is an elementary monomorphism ($\mathfrak{A} \prec \mathfrak{B}$). In Lecture 3 it will be shown that every monomorphism between models of a theory of algebraic type is elementary. Grazie, e buon giorno.

2. Existence of Models

B. g. Today I will describe two approaches to the construction of models, the first via the extended completeness theorem of first order logic, and the second via direct limits. (A structure $\mathfrak{A}$ is said to be a <u>model</u> of a set S of sentences if $\mathfrak{A} \models G$ for every $G \in S$.)

A formula F is a <u>logical consequence</u> of S ($S \vdash F$) if F is among the formulas generated from S as follows: $F \in S$; F is an axiom of first order logic; F is the result of applying some rule of inference of first order logic to $F_1, \ldots, F_n$ when $S \vdash F_i$ $(1 \leq i \leq n)$.

The axioms and rules of first order logic with equality are consonant with common sense, so they need not be listed here. One point worth noting is the finite character of logical consequence: if $S \vdash F$, then $S_0 \vdash F$ for some finite $S_0 \subset S$.

S is logically <u>consistent</u> if no sentence of the form $F \mathbin{\&} \sim F$ is a logical consequence of S.

Theorem 2.1. S is consistent iff S has a model.

Proof. If S has a model, then no contradiction is a logical consequence of S, because every consequence of S is true in every model of S.

Suppose S is consistent. The construction of the model is in two steps. First S is Henkinized by adding new individual constants to the language of S, and new sentences to S that force the new constants to be witnesses. Let $S_0 = S$. S_{n+1} consists of S_n together with all sentences of the form

(1) $$(Ex)F(x) \rightarrow F(\underline{c}_{F(x)}),$$

where F(x) is a formula in the language of S_n, and $\underline{c}_{F(x)}$ is a new individual constant. Let S_∞ be $\bigcup\{S_n \mid n < \omega\}$. The consistency of S_∞ is easily checked by induction on n. For example, suppose S_0 (= S) together with (1) yield a contradiction. Then S_0 yields the negation of (1), and so

$$S_0 \vdash (Ex)F(x) \;\&\; \sim F(\underline{c}_{F(x)}).$$

Since $\underline{c}_{F(x)}$ does not occur in S_0, the derivation of $\sim F(\underline{c}_{F(x)})$ from S_0 is equivalent to the derivation of $\sim F(y)$ from S_0, where y is some variable not occurring in S_0. Since S_0 in no way limits y, the derivation of $\sim F(y)$ from S_0 is equivalent to one of $(y)\sim F(y)$ from S_0. But then

$$S_0 \vdash (Ex)F(x) \;\&\; (y)\sim F(y),$$

an impossibility since S_0 is consistent.

The second step is an application of Zorn's lemma made possible by the finite character of the logical consequence relation $\vdash$. S_∞ is extended to S_∞^*, a maximal consistent set of sentences in the language of S_∞. Note that every sentence or its negation belongs to S_∞.

S_∞^* defines a model $\mathfrak{A}$ of S as follows. With each individual constant $\underline{c}$ of the language $\mathcal{L}_\infty$ of S_∞, associate the equivalence class

$$[\underline{c}] = \{\underline{d} \mid \underline{c} = \underline{d} \in S^*_\infty\}.$$

The universe A of $\mathfrak{A}$ consists of $\{[\underline{c}] \mid \underline{c} \in \mathcal{L}_\infty\}$. The relations, functions and distinguished elements of $\mathfrak{A}$ are defined by:

$$R_i^{\mathfrak{A}}([\underline{c}_1], \ldots, [\underline{c}_n]) \text{ iff } R_i(\underline{c}_1, \ldots, \underline{c}_n) \in S^*_\infty,$$

$$f_j^{\mathfrak{A}}([\underline{c}_1], \ldots, [\underline{c}_n]) = [\underline{c}] \text{ iff } f_j(\underline{c}_1, \ldots, \underline{c}_n) = \underline{c} \in S^*_\infty,$$

$$c_k^{\mathfrak{A}} = [\underline{c}_k].$$

An induction on the complexity of (i.e. number of steps needed to generate) F shows $\mathfrak{A} \models F$ iff $F \in S^*_\infty$, where F is any sentence of $\mathcal{L}_\infty$. It follows $\mathfrak{A}$ is a model of S. □

Theorem 2.1 is the work of K. Gödel, T. Skolem and A. Tarski; the proof given is due to L. Henkin.

A <u>theory</u> T is (according to the definition I prefer) a consistent set of sentences. Thus every theory has a model by 2.1. $T_1 \subset T_2$ means every logical consequence of T_1 is also one of T_2. $T_1 = T_2$ means $T_1 \subset T_2$ and $T_2 \subset T_1$. T is <u>complete</u> if either $T \vdash F$ or $T \vdash \sim F$ for every sentence F in the language of T. By 2.1 T is complete iff all models of T are elementarily equivalent.

Corollary 2.2 (compactness). Suppose S is a set of sentences such that each finite subset of S has a model. Then S has a model.

Countability Proviso: From now on assume every structure has at most countably many relations and functions, and similarly that every language has at most countably many relation symbols and function symbols.

Corollary 2.3 (upward Skolem-Löwenheim). Suppose $\mathfrak{A}$ is an infinite structure. Then $\mathfrak{A}$ has a proper elementary extension of cardinality κ for each $\kappa \geq$ cardinality of $\mathfrak{A}$.

Proof. Let T be the complete theory of $<\mathfrak{A}, a>_{a \in A}$, that is the set of all sentences of $\mathcal{L}_{\tau A}$ true in $\mathfrak{A}$, where τ is the similarity type of $\mathfrak{A}$. The models of T coincide with the elementary extensions of $\mathfrak{A}$. Let $\{\underline{c}_\delta \mid \delta < \kappa\}$ be a set of individual constants, none of which occur in T, and let S be

$$T \cup \{\underline{c}_0 \neq \underline{a} \mid a \in A\}$$
$$\cup \{\underline{c}_\gamma \neq \underline{c}_\delta \mid \gamma < \delta < \kappa\}.$$

S is consistent because $\mathfrak{A}$ is infinite. Any model of S is a proper elementary extension of $\mathfrak{A}$ of cardinality at least κ. The details of Theorem 2.1 imply S has a model of cardinality at most κ, since the cardinality of the language of S is at most κ. □

A directed set $<D, \leq>$ consists of a set D with a partial ordering $\leq$ such that for any $i, j \in D$, there is a $k \in D$ such that $i, j \leq k$. A <u>direct system</u> $\{\mathfrak{A}_i, m_{ij}\}$ of structures and monomorphisms consists of a directed set $<D, \leq>$, a family $\{\mathfrak{A}_i \mid i \in D\}$ of structures, and a family $\{m_{ij} : \mathfrak{A}_i \to \mathfrak{A}_j \mid i \leq j \in D\}$ of monomorphisms such that $m_{ii} : \mathfrak{A}_i \to \mathfrak{A}_i$ is the identity monomorphism and $m_{ik} = m_{jk} m_{ij}$ whenever $i \leq j \leq k$.

Let $A = \bigcup\{A_i \times \{i\} \mid i \in D\}$. Call $\langle a, i\rangle$ equivalent to $\langle b, j\rangle$ if $m_{ik}a = m_{jk}b$ for some $k \in D$. Let $[a, i]$ be the equivalence class of $\langle a, i\rangle$, and let A_∞ be the set of all such equivalence classes. The direct limit of $\{\mathcal{A}_i, m_{ij}\}$, denoted by $\lim\limits_{\rightarrow} \mathcal{A}_i$ or $\mathcal{A}_\infty$, is a structure whose universe is A_∞. The relation $R^{\mathcal{A}_\infty}([a_1, i_1], \ldots, [a_n, i_n])$ holds iff $R^{\mathcal{A}_k}(m_{i_1 k}a_1, \ldots, m_{i_n k}a_n)$ holds for some k such that $i_t \leq k$ for $1 \leq t \leq n$. The functions and distinguished elements of $\mathcal{A}_\infty$ are defined similarly. The monomorphism $m_{i\infty} : \mathcal{A}_i \rightarrow \mathcal{A}_\infty$ is given by $m_{i\infty}a = [a, i]$. Note that $m_{j\infty}m_{ij} = m_{i\infty}$ if $i \leq j$. Strictly speaking, the direct limit of $\{\mathcal{A}_i, m_{ij}\}$ consists of $\mathcal{A}_\infty$ and $\{m_{i\infty} : \mathcal{A}_i \rightarrow \mathcal{A}_\infty\}$.

Theorem 2.4 (A. Tarski, R. Vaught). Suppose $\{\mathcal{A}_i, m_{ij}\}$ is a direct system of structures and elementary monomorphisms. Then $m_{i\infty} : \mathcal{A}_i \rightarrow \mathcal{A}_\infty$ is elementary for all i.

Proof. For all i show by induction on the complexity of formulas that

$$\mathcal{A}_i \models F(\underline{a}_1, \ldots, \underline{a}_n) \text{ iff } \mathcal{A}_\infty \models F(\underline{m_{i\infty}}a_1, \ldots, \underline{m_{i\infty}}a_n)$$

for all F and all $a_1, \ldots, a_n \in A_i$. For example, suppose $\mathcal{A}_\infty \models (Ex)G(x, \underline{m_{i\infty}}a)$, where $a \in A_i$. Then

$$\mathcal{A}_\infty \models G(\underline{m_{j\infty}}b, \underline{m_{i\infty}}a)$$

for some $j \in D$ and $b \in A_j$. Choose k so that $i, j \leq k$. Then

$$\mathcal{A}_\infty \models G(\underline{m_{k\infty}m_{jk}}b, \underline{m_{k\infty}m_{ik}}a).$$

G is less complex than $(Ex)G$, so $\mathcal{A}_k \models G(\underline{m_{jk}}b, \underline{m_{ik}}a)$. Hence $\mathcal{A}_k \models (Ex)G(x, \underline{m_{ik}}a)$. But then $\mathcal{A}_i \models (Ex)G(x, \underline{a})$, since m_{ik} is elementary. □

Suppose γ is an ordinal, and $\{\mathcal{A}_\alpha | \alpha < \gamma\}$ is a family of structures such that $\mathcal{A}_\alpha \subset \mathcal{A}_\beta$ when $\alpha < \beta < \gamma$. $\{\mathcal{A}_\alpha | \alpha < \gamma\}$ is said to be a chain of length γ. It can be viewed as a direct system $\{\mathcal{A}_\alpha, i_{\alpha\beta}\}$, where $\alpha < \beta$ and $i_{\alpha\beta} : \mathcal{A}_\alpha \subset \mathcal{A}_\beta$ is the inclusion monomorphism. The union of the chain $\{\mathcal{A}_\alpha | \alpha < \gamma\}$ is defined to be the direct limit of $\{\mathcal{A}_\alpha, i_{\alpha\beta}\}$. A chain is said to be elementary if all of the associated inclusion monomorphisms are elementary.

Corollary 2.5 (elementary chain principle). The union of an elementary chain is an elementary extension of every member of the chain.

The extended completeness theorem serves to erect elementary chains (as in the proof of 2.3) whose limits are of interest thanks to the elementary chain principle. G. e b. g.

3. Model Completions of Theories

B. g. Today I will discuss A. Robinson's notion of model completion of a theory, his first and possibly most successful attempt to generalize algebraic closedness. My prime example will be the theory of algebraically closed fields viewed as the model completion of the theory of fields, and I will use it to illustrate how some of the direct arguments of elimination theory can be replaced by indirect arguments via the extended completeness theorem of first order logic.

Let T and T_1 be theories in the same language. T_1 is a <u>model completion</u> of T if for all $\mathcal{A}$, $\mathcal{B}$ and $\mathcal{C}$:

(i) if $\mathcal{A} \models T_1$, then $\mathcal{A} \models T$;

(ii) if $\mathcal{A} \models T$, then there exists a $\mathcal{B} \supset \mathcal{A}$ such that $\mathcal{B} \models T_1$;

(iii) if $\mathcal{A} \models T$, $\mathcal{A} \subset \mathcal{B}_i$ and $\mathcal{B}_i \models T_1$ $(i = 1,2)$, then $\langle \mathcal{B}_1, a \rangle_{a \in A} \equiv \langle \mathcal{B}_2, a \rangle_{a \in A}$.

Theorem 3.1 (A. Robinson). If T_1 and T_2 are model completions of T, then $T_1 = T_2$.

Proof. By 2.1 and the symmetry of the situation, it is enough to show an arbitrary model $\mathcal{A}$ of T_1 is also a model of T_2. Define a chain $\{\mathcal{A}_n \mid n < \omega\}$ such that $\mathcal{A}_0 = \mathcal{A}$, $\mathcal{A}_{2n} \models T_1$ and $\mathcal{A}_{2n+1} \models T_2$. $\{\mathcal{A}_{2n} \mid n < \omega\}$ is an elementary chain, so $\mathcal{A}_0 \prec \mathcal{A}_\infty$ by 2.5. Similarly, $\mathcal{A}_1 \prec \mathcal{A}_\infty$, so $\mathcal{A}_0 \prec \mathcal{A}_1$ by 1.1(ii). □

The theory of algebraically closed fields (ACF) is the theory of fields (TF) augmented by

$$(y_1)\ldots(y_n)(Ex)[x^n + y_1 x^{n-1} + \ldots + y_{n-1}x + y_n = 0]$$

for each $n > 0$.

Theorem 3.2 (A. Robinson). ACF is the model completion of TF.

Proof. Suppose $\mathcal{A}$ is a field with algebraically closed extensions $\mathcal{B}_1$ and $\mathcal{B}_2$. To see $\langle \mathcal{B}_1, a \rangle_{a \in A} \equiv \langle \mathcal{B}_2, a \rangle_{a \in A}$, use 2.3 to obtain $\mathcal{C}_i$, an elementary extension of $\mathcal{B}_i$ of cardinality κ, where $\kappa > \operatorname{card} \mathcal{A}$. The identity map on $\mathcal{A}$ can be extended to an isomorphism between $\mathcal{C}_1$ and $\mathcal{C}_2$, since the latter are algebraically closed fields of the same transcendence rank over $\mathcal{A}$. Now apply 1.1. □

A theory T admits <u>elimination of quantifiers</u> if for each formula F (in the language of T), there is a formula G without quantifiers such that $T \vdash F \leftrightarrow G$. A formula is <u>universal</u> if it is of the form $(x_1)\ldots(x_n)H$, where $n \geq 0$ and H has no quantifiers. A theory V is said to be <u>universal</u> if there exists a theory W such that V = W and every member of W is a universal sentence. TF is a typical universal theory.

Theorem 3.3 (A. Robinson). If T is the model completion of a universal theory, then T admits elimination of quantifiers.

Proof. Suppose T is the model completion of V, where V is universal. Let $\mathcal{C}$ be any substructure of any model of T with the intent of showing $T \cup D\mathcal{C}$ is complete. ($D\mathcal{C}$ is the <u>diagram</u> of $\mathcal{C}$, the set of all atomic sentences or negations of atomic sentences true in $\langle \mathcal{C}, c \rangle_{c \in C}$.) Every universal sentence provable in T must be true in $\mathcal{C}$; hence $\mathcal{C} \models V$. By clause (iii) of the definition of model completion, all models of $T \cup D\mathcal{C}$ are elementarily equivalent.

Now let F(x) be a formula in the language of T, and let S be the following set of sentences: T; $F(\underline{c})$, where $\underline{c}$ does not occur in T; and $\sim K(\underline{c})$, where K(x) is any quantifierless formula such that

$T \vdash K(x) \rightarrow F(x)$. To see S is inconsistent, assume S has a model $\mathcal{A}$. Let $\mathcal{C}$ be the least substructure of $\mathcal{A}$ with c as a member. Then $T \cup D\mathcal{C} \vdash F(\underline{c})$, since $T \cup D\mathcal{C}$ is complete, $\mathcal{A}$ is a model of $T \cup D\mathcal{C}$, and $\mathcal{A} \models F(\underline{c})$. But then $T \vdash K(\underline{c}) \rightarrow F(\underline{c})$, where $K(\underline{c})$ is finitely much of $D\mathcal{C}$; clearly $K(\underline{c})$ is quantifierless and $\mathcal{A} \models K(\underline{c})$. Since $\underline{c}$ does not occur in T, it follows that $T \vdash K(x) \rightarrow F(x)$, a contradiction because then the definition of S requires $\mathcal{A} \models \sim K(\underline{c})$.

The inconsistency of S implies $T \vdash F(x) \rightarrow J(x)$, where $J(x)$ is the disjunction of finitely many quantifierless $K(x)$'s, each with the property that $T \vdash K(x) \rightarrow F(x)$. □

Corollary 3.4 (A. Tarski, A. Robinson). The theory of algebraically closed fields admits elimination of quantifiers.

Two related consequences of 3.4 concern algebraic sets and solvability of finite systems of polynomial equations. An n-dimensional, complex <u>algebraic set</u> consists of all complex solutions of some finite system of polynomial equations in n variables. By 3.4 the projection of an n-dimensional, complex algebraic set on m-dimensional complex space is a finite intersection of finite unions of m-dimensional, complex algebraic sets and their complements.

Let S be a finite system of polynomial equations and inequations in several variables with coefficients $c_1, \ldots, c_n$. The assertion S has a solution is expressible by some first order sentence $F(\underline{c}_1, \ldots, \underline{c}_n)$. By 3.4 there is a quantifierless formula $H(x_1, \ldots, x_n)$ such that

(1) $$\mathrm{ACF} \vdash F(x_1,\ldots,x_n) \leftrightarrow H(x_1,\ldots,x_n).$$

If $\mathfrak{A}$ is an algebraically closed field containing $c_1,\ldots,c_n$, then S has a solution in $\mathfrak{A}$ iff $\mathfrak{A} \models H(\underline{c}_1,\ldots,\underline{c}_n)$. H constitutes an algebraic criterion for the solvability of S, because computing the truth value of $H(\underline{c}_1,\ldots,\underline{c}_n)$ is equivalent to evaluating finitely many polynomials (with coefficients in the prime subfield) at $\langle c_1,\ldots,c_n\rangle$. The existence of H was established in this lecture by little more than the extended completeness theorem of first order logic and the uniqueness of an algebraically closed field of given characteristic and dimension. To compute H from S in an efficient fashion, one must appeal to Kronecker's elimination theory. A crude procedure for finding an H that satisfies (1) is to enumerate all first order proofs based on the axioms of ACF. The procedure is effective, because the usual axioms for ACF form a recursive set. G. e b. g.

4. Isomorphism Types of Simple Extensions

B. g. Today I wish to talk about the leading role played by the notion of simply generated extension of a substructure of a model in the study of theories of algebraic type. Consider the example of ACF_0. A substructure $\mathfrak{A}$ of a model of ACF_0 is nothing more nor less than a field of characteristic 0, since $\mathfrak{A}$ must include the distinguished elements 0 and 1, and must be closed with respect to addition, multiplication, additive inverse and multiplicative inverse. The simple extensions of $\mathcal{A}$ correspond to roots of polynomials irreducible over $\mathcal{A}$, save for the unique simple transcendental extension of $\mathcal{A}$. The properties of ACF_0 established in the previous lectures were immediate consequences of the nature of the simple extensions of fields. Recall that the axioms of ACF_0 allude only to polynomials in one variable, a restriction that is no accident according to the following theorem.

Theorem 4.1 (L. Blum [1]). Let V be the model completion of some universal theory. Then there exists a theory W such that V = W and every member of W is of the form $(y_1)\ldots(y_n)(Ex)F(y_1,\ldots,y_n,x)$, where $F(y_1,\ldots,y_n,x)$ is quantifierless.

Blum's proof of 4.1 is a typical application of saturated models (cf. [2], p. 89).

From now on assume T is a theory that admits elimination of quantifiers. Let $\mathcal{K}(T)$ be the category of all substructures of all models of T. Suppose $\mathcal{A} \subset \mathcal{B} \in \mathcal{K}(T)$. $\mathcal{B}$ is said to be a simple extension of $\mathcal{A}$ if there is a $b \in B$ such that $\mathcal{B}$ is the least substructure of $\mathcal{B}$ whose universe contains $A \cup \{b\}$, in symbols $\mathcal{A}(b) = \mathcal{B}$. $\mathcal{A}(b)$ and $\mathcal{A}(c)$ are isomorphic over $\mathcal{A}$ if there exists an isomorphism $f : \mathcal{A}(b) \overset{\approx}{\to} \mathcal{A}(c)$ such that $f|A = 1_A$ and $fb = c$. Such an f is unique since each member of $\mathcal{A}(b)$ is named by a term $t(b)$, where $t(x)$ is a term in the language of $T \cup D\mathcal{A}$.

Since T admits elimination of quantifiers, the theory $T \cup D\mathcal{A}$ is complete, that is all models of T containing $\mathcal{A}$ satisfy the same sentences of the language of T with constants added to name the elements of $\mathcal{A}$. The isomorphism types of simple extensions of $\mathcal{A}$ correspond to the 1-types of $T \cup D\mathcal{A}$. Let $F_1(T \cup D\mathcal{A})$ be the set of all formulas in the language of $T \cup D\mathcal{A}$ of the form $F(x)$. Call $F(x)$ consistent if $T \cup D\mathcal{A} \vdash (Ex)F(x)$, and call a set $S \subset F_1(T \cup D\mathcal{A})$ consistent if the conjunction of any finite number of members of S is consistent. A 1-type p is a maximal consistent subset of $F_1(T \cup D\mathcal{A})$. b is said to realize p if for every $F(x) \in p$, $F(\underline{b})$ is true in every model of T which extends $\mathcal{A}(b)$.

Let $S\mathcal{A}$ be the set of all 1-types of $T \cup D\mathcal{A}$. If $p \in S\mathcal{A}$, then the compactness theorem (2.2) implies there exists an $\mathcal{A}(b)$ such that b realizes p. Conversely, for each $\mathcal{A}(b)$ the set of all $F(x)$ such that $F(\underline{b})$ holds in any model of T extending $\mathcal{A}(b)$ is a 1-type of $S\mathcal{A}$. Furthermore, b and c are isomorphic over $\mathcal{A}$ iff they give rise to the same 1-type.

Suppose $i : \mathcal{A} \to \mathcal{P}$ is an extension of $\mathcal{A}$ to a model $\mathcal{P}$ of T. $\mathcal{P}$ is said to be a prime model extension of $\mathcal{A}$ if for every extension $f : \mathcal{A} \to \mathcal{C}$ of $\mathcal{A}$ to a model $\mathcal{C}$ of T, there exists a $g : \mathcal{C} \to \mathcal{P}$ such that $i = gf$. For example, the algebraic closure of a field $\mathcal{A}$ of characteristic 0 is a prime model extension of $\mathcal{A}$ in the context of algebraically closed fields of characteristic 0. My seventh lecture will be devoted to showing that for theories of algebraic type, every substructure of every model has a unique prime model extension. Both existence and uniqueness will be derived from the Morley analysis of $S\mathcal{A}$ construed as a compact Hausdorff space whose clopen sets form a base for its topology. A typical basic open subset of $S\mathcal{A}$, denoted by $U_{F(x)}$, is $\{p \mid F(x) \in p\}$, where $F(x) \in F_1(T \cup D\mathcal{A})$.

Let $f : \mathcal{A} \to \mathcal{B}$ belong to $\mathcal{H}(T)$. Define $Sf : S\mathcal{B} \to S\mathcal{A}$ by: $F(\underline{a}_1, \ldots, \underline{a}_n, x) \in (Sf)q$ iff $F(\underline{fa}_1, \ldots, \underline{fa}_n, x) \in q$, where $q \in S\mathcal{B}$. Sf is continuous and onto. Let $\mathcal{H}$ be the category of compact Hausdorff spaces and continuous onto maps. $S : \mathcal{H}(T) \to \mathcal{H}$ is the contravariant functor which assigns to each structure $\mathcal{A}$ of $\mathcal{H}(T)$ the compact Hausdorff space $S\mathcal{A}$, and to each monomorphism $f : \mathcal{A} \to \mathcal{B}$ of $\mathcal{H}(T)$ the continuous onto map $Sf : S\mathcal{B} \to S\mathcal{A}$.

Direct systems in $\mathcal{H}(T)$ are transformed by S into inverse systems of $\mathcal{H}$. An <u>inverse system</u> in $\mathcal{H}$ consists of a directed set $\langle D, \leq \rangle$, a family $\{X_i \mid i \in D\}$ of spaces of $\mathcal{H}$, and a family $\{f_{ji} : X_j \to X_i \mid i \leq j\}$ of maps of $\mathcal{H}$ such that: f_{ii} is the identity map on X_i; and $f_{ki} = f_{ji} f_{kj}$ whenever $i \leq j \leq k$. To form the inverse limit of $\{X_i, f_{ji}\}$, let X_∞ be

$$\{x \mid x \in \prod_{i \in D} X_i \ \& \ f_{ji} x_j = x_i \ \text{ for all } \ j \geq i\},$$

where x_i is the i-th coordinate of x. Assign the product topology to $\prod_i X_i$. Then X_∞ is a closed, hence compact, subspace of $\prod_i X_i$. Thus $X_\infty \in \mathcal{H}$. Define $f_{\infty i} : X_\infty \to X_i$ by $f_{\infty i} x = x_i$. Clearly $f_{\infty i}$ is continuous; it is onto by virtue of the compactness of $\prod_i X_i$.

To see that $\{X_\infty, f_{\infty i}\}$ is the inverse limit of $\{X_i, f_{ji}\}$ in the sense of category theory, suppose Y is a compact Hausdorff space and $\{g_i : Y \to X_i\}$ is a family of continuous onto maps such that $f_{ji} g_j = g_i$

for all $j \geq i$. Define $g : Y \to X_\infty$ by $(gy)_i = g_i y$ for each $y \in Y$. Clearly $f_{\infty i} g = g_i$ for all i. The "universal property" of the inverse limit is merely the uniqueness of g: if $h : Y \to X_\infty$ is such that $f_{\infty i} h = g_i$ for all i, then $h = g$.

Only two definitions separate us from the next lemma. Let $G : \mathcal{K}(T) \to \mathcal{H}$ be a contravariant functor. Thus if $\{\mathcal{A}_i, f_{ij}\}$ is a direct system of substructures of models of T and monomorphisms, then $\{G\mathcal{A}_i, Gf_{ij}\}$ is an inverse system of compact Hausdorff spaces and onto maps. G <u>preserves limits</u> if the following holds: suppose the direct limit of $\{\mathcal{A}_i, f_{ij}\}$ consists of $\mathcal{A}_\infty$ and $\{f_{i\infty} : \mathcal{A}_i \to \mathcal{A}_\infty\}$, and the inverse limit of $\{G\mathcal{A}_i, Gf_{ij}\}$ consists of $\varprojlim G\mathcal{A}_i$ and $\{g_{\infty i} : \varprojlim G\mathcal{A}_i \to G\mathcal{A}_i\}$; let $g : G\mathcal{A}_\infty \to \varprojlim G\mathcal{A}_i$ be the unique map such that $g_{\infty i} g = Gf_{i\infty}$ for all i; then g is a homeomorphism.

A category $\mathcal{K}$ admits filtrations with amalgamation if for each pair $j : \mathcal{A} \to \mathcal{B}$, $k : \mathcal{A} \to \mathcal{C}$ in $\mathcal{K}$, there exists a pair $f : \mathcal{B} \to \mathcal{D}$, $g : \mathcal{C} \to \mathcal{D}$ in $\mathcal{K}$ such that $fj = gk$.

Lemma 4.1. (i) $\mathcal{K}(T)$ admits filtrations with amalgamation. (ii) The contravariant functor $S : \mathcal{K}(T) \to \mathcal{H}$ preserves limits.

Proof. (i) Suppose $\mathcal{B}$ and $\mathcal{C}$ are models of T with a common substructure $\mathcal{A}$. For a model $\mathcal{D}$ of T which extends both $\mathcal{B}$ and $\mathcal{C}$, take any model of the following set V of sentences: T; all sentences of $\mathcal{L}_{\tau B}$ true in $\mathcal{B}$, where $\mathcal{L}_\tau$ is the language underlying $\mathcal{B}$; and all sentences of $\mathcal{L}_{\tau C}$ true in $\mathcal{C}$. The consistency of V follows from the elementary equivalence of $\langle \mathcal{B}, a\rangle_{a \in A}$ and $\langle \mathcal{C}, a\rangle_{a \in A}$, which is implied by the assumption T admits elimination of quantifiers.

(ii) Let $\{\mathcal{A}_i, f_{ij}\}$ be a direct system of $\mathcal{K}(T)$, and let $\varprojlim S\mathcal{A}_i$, together with $\{g_{\infty i} : \varprojlim S\mathcal{A}_i \to S\mathcal{A}_i\}$, be the inverse limit of $\{S\mathcal{A}_i, Sf_{ij}\}$. Thus $(Sf_{ij})g_{\infty j} = g_i$ when $i \leq j$. Let $\mathcal{A}_\infty$, together with $\{f_{i\infty} : \mathcal{A}_i \to \mathcal{A}_\infty\}$, be the direct limit of $\{\mathcal{A}_i, f_{ij}\}$. There exists a unique $g : S\mathcal{A}_\infty \to \varprojlim S\mathcal{A}_i$ such that $g_{\infty i} g = Sf_{i\infty}$ for all i. If g is one-one, then g is a homeomorphism, because every one-one, continuous onto map between compact Hausdorff spaces is a homeomorphism. To check g is one-one, suppose $gp = gq$. Then $g_{\infty i} gp = g_{\infty i} gq$, and so $Sf_{i\infty} p = Sf_{i\infty} q$ for all i. But then $p = q$, since any difference between p and q would be manifested by a formula belonging to $F_1(T \cup D\mathcal{A}_i)$ for some i.

□

Tomorrow I will develop the Morley derivative DS of S, and use it to define the concept of quasi-total transcendality. G. e b. g.

5. Quasi-Total Transcendality

B. g. Today I will finish unpacking the notion of theory of algebraic type by giving a definition of quasi-totally transcendental theory, a concept due to L. Blum [1]. The simplest example of the concept, but not the most misleading, is the theory of real closed fields (RCF). The axioms of RCF are those of TF plus:

$$(x) \sim (x < x)$$

$$(x)(y)(z)[x < y \,\&\, y < z \rightarrow x < z]$$

$(x)(y)[x < y$ [illegible] $x = y \vee y < x]$

$(x)(y)[0 <$ [illegible] $0 < y \rightarrow 0 < x \cdot y]$

$$(x)(y)(z)[x < y \rightarrow x + z < y + z]$$

$$(x)(Ey)[0 < x \rightarrow x = y \cdot y]$$

$$(x_1) \ldots (x_n)(Ey)[y^n + x_1 y^{n-1} + \ldots + x_n = 0] \quad (n \text{ odd}).$$

RCF is the model completion of OF, the theory of ordered fields. (OF is TF plus the first five of the above axioms.) Hence by 3.3 RCF admits elimination of quantifiers, first proved by A. Tarski via his improvements of Sturm's algorithm [3]. A less direct proof via the extended completeness theorem in the setting of saturated models (as in [2], p. 92) requires only one algebraic fact: every simple algebraic extension of an ordered field $\mathfrak{A}$ (extension in the sense of ordered fields) can be found inside every real closed extension of $\mathfrak{A}$. More generally L. Blum [1] (or [2], p. 89) has given a necessary and sufficient condition for a theory T_2 to be the model completion of a universal theory T_1, namely, every simple extension of any model $\mathfrak{A}$ of T_1 can be found inside every sufficiently saturated model of T_2 which extends $\mathfrak{A}$.

The simple extensions of an ordered field $\mathfrak{A}$ are of two kinds, algebraic and transcendental. A simple extension $\mathfrak{A}(b)$ of $\mathfrak{A}$ is completely specified by the polynomial equalities and inequalities over $\mathfrak{A}$ satisfied by b. (Note: a polynomial over $\mathfrak{A}$ can have <u>two</u> roots in the same cut in the ordering of $\mathfrak{A}$.) Any simple extension $\mathfrak{A}(b)$ can be well approximated by a simple algebraic extension $\mathfrak{A}(b^*)$ in the sense that b^* realizes a preassigned finite portion of the specification of $\mathcal{A}(b)$. In short, the simple algebraic extensions of $\mathcal{A}$ constitute a dense subset of $S\mathcal{A}$. Morley's notion of ranked is a generalization of algebraic, and a theory T is said to be quasi-totally transcendental if the ranked elements of $S\mathcal{A}$ are dense in $S\mathcal{A}$ for every $\mathcal{A} \in \mathcal{H}(T)$.

As in Lemma 4.1 assume T admits elimination of quantifiers. The Morley derivative DS of $S : \mathcal{H}(T) \to \mathcal{H}$ is defined as follows. For each $\mathcal{A} \in \mathcal{H}(T)$, $p \in DS\mathcal{A}$ iff $p \in S\mathcal{A}$ and there exists a monomorphism $f : \mathcal{A} \to \mathcal{B}$ in $\mathcal{H}(T)$ such that $(Sf)^{-1}p$ has a limit point in $S\mathcal{B}$. For each $f : \mathcal{A} \to \mathcal{B}$ in $\mathcal{H}(T)$, DSf is the restriction of Sf to $DS\mathcal{B}$. To see that DSf maps $DS\mathcal{B}$ into $DS\mathcal{A}$, fix $q \in DS\mathcal{B}$. For some $g : \mathcal{B} \to \mathcal{C}$, $(Sg)^{-1}q$ contains a limit point, hence $(Sgf)^{-1}Sfq$ contains a limit point, and so $DSfq \in DS\mathcal{A}$.

Now fix $p \in DS\mathcal{A}$ to find a $q \in DS\mathcal{B}$ such that $DSfq = p$. Choose $h : \mathcal{A} \to \mathcal{C}$ so that $(Sh)^{-1}p$ contains a limit point. By 4.1(i) there is a $\mathcal{D}$ such that $S\mathcal{D}$ projects onto both $S\mathcal{B}$ and $S\mathcal{C}$. The pre-image of $(Sh)^{-1}p$ in $S\mathcal{D}$ contains some limit point r. The image of r in $S\mathcal{B}$ is the desired q.

Thus $DS : \mathcal{H}(T) \to \mathcal{H}$ is a contravariant functor. In addition DS preserves limits by essentially the same argument given for 4.1(ii) but making use of the fact that S preserves limits.

For each ordinal α, the α-th Morley derivative of S, denoted by $D^{\alpha}S$, is defined by recursion: $D^0S = S$; $D^{\alpha+1}S = D(D^{\alpha}S)$; and $D^{\lambda}S\mathcal{A} = \bigcap \{D^{\alpha}S\mathcal{A} \mid \alpha < \lambda\}$ when λ is a limit ordinal. Of course $D^{\lambda}Sf$ is the restriction of Sf to $D^\lambda S\mathcal{B}$, if $f : \mathcal{A} \to \mathcal{B}$ is in $\mathcal{H}(T)$.

Lemma 5.1. For each α, $D^\alpha S : \mathcal{H}(T) \to \mathcal{H}$ is a contravariant functor that preserves limits.

The proof of 5.1 is by induction on α. The details of the definitions of S and $\mathcal{H}(T)$ are irrelevant; one needs only that S preserves limits and $\mathcal{H}(T)$ admits filtrations with amalgamation in order to argue as in the proof of 4.1(ii) (cf. [2], p. 176).

Suppose there is an α (necessarily unique) such that $p \in D^\alpha S\mathcal{A} - D^{\alpha+1}S\mathcal{A}$. Then p is said to be a (<u>Morley</u>) <u>ranked</u> point of $S\mathcal{A}$ of rank α.

Lemma 5.2 (Rank Rule). Suppose p is a ranked point of $S\mathcal{A}$ and $f : \mathcal{A} \to \mathcal{B}$ belongs to $\mathcal{H}(T)$. If $Sfq = p$, then q is ranked and rank $q \leq$ rank p.

The rank rule is an immediate consequence of the fact that Sf maps $D^{\alpha+1}S\mathcal{B}$ <u>onto</u> $D^{\alpha+1}S\mathcal{A}$ for all α.

Proposition 5.3. Suppose $p \in S\mathcal{A} - DS\mathcal{A}$. Then there exists a positive integer n such that for all $f : \mathcal{A} \to \mathcal{B}$ in $\mathcal{H}(T)$, the cardinality of $(Sf)^{-1}p$ is at most n.

Proof. Suppose not. Then there is a chain $f_n : \mathcal{A}_n \to \mathcal{A}_{n+1}$ $(n < \omega)$ such that $\mathcal{A}_0 = \mathcal{A}$ and $(Sf_n \ldots f_1 f_0)^{-1}p$ has cardinality at least n, because $\mathcal{H}(T)$ admits filtrations with amalgamation. But then the pre-image of p in $\mathcal{A}_\infty$ is infinite and so must contain a limit point. □

The proof of 5.3 applies to every limit preserving functor from $\mathcal{K}(T)$ into $\mathcal{H}$, and hence to $D^\alpha S$ for all α by 5.1. If $p \in S\mathcal{A}$ has rank α, then the maximum value of the cardinality of $(D^\alpha Sf)^{-1}p$, for all $f : \mathcal{A} \to \mathcal{B}$ in $\mathcal{K}(T)$, is called the <u>degree</u> of p.

Proposition 5.4 (Degree Rule). If $p \in S\mathcal{A}$ is ranked and $f : \mathcal{A} \to \mathcal{B}$ belongs to $\mathcal{K}(T)$, then

$$\deg p = \Sigma\{\deg q \mid Sfq = p \ \& \ \operatorname{rank} q = \operatorname{rank} p\}.$$

The degree rule follows easily from the fact that $\mathcal{K}(T)$ admits filtrations with amalgamation.

Suppose T is quasi-totally transcendental. The <u>density number</u> of T, denoted by d_T and defined by Blum [1], is the least α such that for all $\mathcal{A} \in \mathcal{K}(T)$, the ranked points of $S\mathcal{A}$ of rank less than α are dense in $S\mathcal{A}$. Thus $d_{RCF} = 1$, as indicated earlier in this lecture, and $d_{DCF_0} = \omega$ (DCF_0 is the theory differentially closed fields of characteristic 0), as will be noted in the concluding lecture. It can be shown that d_T is always countable [4] despite the fact there exists a quasi-totally transcendental T and an $\mathcal{A} \in \mathcal{K}(T)$ such that the ranked points of $S\mathcal{A}$ do not have their ranks bounded by a countable ordinal.

I hope that I have not made Morley rank more mysterious than it is. If I have, perhaps the following two observations will help. First, the Morley analysis of $S\mathcal{A}$ does not require all of $\mathcal{K}(T)$ but only those structures in $\mathcal{K}(T)$ of the same cardinality as $\mathfrak{A}$. Second, the Morley analysis of $S\mathfrak{A}$ is closely related to the Cantor-Bendixson analysis of $S\mathfrak{A}$. In particular there exists a $\mathfrak{B} \supset \mathfrak{A}$ such that $\mathfrak{B}$ has the same cardinality as $\mathfrak{A}$, and such that the Cantor-Bendixson assignment of rank to elements of $S\mathfrak{B}$ is the same as that of Morley.

In the next two lectures I will use Morley rank to prove theorems about models of quasi-totally transcendental theories by induction on the rank of the 1-types realized in those models. G. e b. g.

6. Indiscernibles and Prime Model Extensions

B. g. Today I will show how Morley's notion of rank can be used to study sets of indiscernibles in models of quasi-totally transcendental theories. Sets of indiscernibles very much resemble sets of algebraically independent field elements, and will serve as "transcendence bases" in the proof of uniqueness of prime model extensions lying in wait in tomorrow's lecture.

The elements of I are <u>indiscernible</u> in $\mathfrak{A}$ if $I \subset A$ and $\mathfrak{A} \models F(\underline{i}_1, \ldots, \underline{i}_n) \leftrightarrow F(\underline{j}_1, \ldots, \underline{j}_n)$ whenever $\{i_1, \ldots, i_n\}$ and $\{j_1, \ldots, j_n\}$ are n-element subsets of I and $F(x_1, \ldots, x_n)$ is a formula in the language underlying $\mathfrak{A}$. If $Y \subset A$, then I is said to be indiscernible over Y in $\mathfrak{A}$ if the elements of I are indiscernible in $\langle \mathfrak{A}, y \rangle_{y \in Y}$.

Any set of algebraically independent elements in any algebraically closed field K is indiscernible over the prime subfield of K.

As in the last lecture, assume T admits elimination of quantifiers, and consider $\mathcal{A} \in \mathcal{K}(T)$, $Y \subset A$ and a wellordered subset $\{a_\delta \mid \delta < \alpha\}$ of A. For each $\delta < \alpha$, let p_δ be the 1-type realized by a_δ over $Y \cup \{a_\gamma \mid \gamma < \delta\}$. (To be precise, $p_\delta \in S\mathcal{B}_\delta$, where $\mathcal{B}_\delta$ is the least substructure of $\mathcal{A}$ whose universe contains $Y \cup \{a_\gamma \mid \gamma < \delta\}$, but there is no harm in ignoring the distinction between the two.) Suppose p_0 is ranked. Then $\{a_\delta \mid \delta < \alpha\}$ is said to be a Morley sequence over Y in $\mathcal{A}$ if p_δ is a pre-image of p_γ of the same rank and degree as p_γ when $0 \leq \gamma < \delta < \alpha$. ($Sfp_\delta = p_\gamma$, where $f : \mathcal{B}_\gamma \subset \mathcal{B}_\delta$.) Any sequence of algebraically independent field elements is a Morley sequence over any algebraic extension of the prime subfield with p_δ of rank 1 and degree 1 for all δ.

Lemma 6.1. Every infinite Morley sequence over Y in $\mathcal{A}$ is indiscernible over Y in $\mathcal{A}$.

Proof. The sequence $\{a_\delta \mid \delta < \alpha\}$ is said to be order indiscernible over Y in $\mathcal{A}$ if

$$\mathcal{A} \models F(\underline{a}_{\delta_1}, \ldots, \underline{a}_{\delta_n}) \leftrightarrow F(\underline{a}_{\gamma_1}, \ldots, \underline{a}_{\gamma_n})$$

whenever $\delta_1 < \ldots < \delta_n$, $\gamma_1 < \ldots < \gamma_n$ and $F(x_1, \ldots, x_n)$ is a formula in the language underlying $\langle \mathcal{A}, y \rangle_{y \in Y}$. A compactness argument having little to do with rank shows any infinite set of order indiscernibles over Y, all of whose members realize the same ranked 1-type (p_0), is a set of indiscernibles over Y (cf. [2], p. 229). The order indiscernibility is established by induction on n. When $n = 1$, all is well since all the a_δ's satisfy $p_0 \in SY$. Fix $n > 1$. By induction there is an isomorphism

$$j : Y(a_{\delta_1}, \ldots, a_{\delta_{n-1}}) \overset{\approx}{\rightarrow} Y(a_{\gamma_1}, \ldots, a_{\gamma_{n-1}}).$$

Let p be the type realized by a_{δ_n} over $Y(a_{\delta_1}, \ldots, a_{\delta_{n-1}})$, and q the type realized by a_{γ_n} over $Y(a_{\gamma_1}, \ldots, a_{\gamma_{n-1}})$. It suffices to prove $(Sj)q = p$.

By the rank rule (5.2), $\operatorname{rank} p_0 \geq \operatorname{rank} p \geq \operatorname{rank} p_{\delta_n}$. Hence $\operatorname{rank} p = \operatorname{rank} p_0$, since $\{a_\delta\}$ is a Morley sequence. A similar approach via the degree rule (5.4) shows that $\deg p = \deg p_0$, and that p is the <u>unique</u> pre-image of p_0 in $SY(a_{\delta_1}, \ldots, a_{\delta_{n-1}})$ of the same rank and degree as p_0. But the same holds for q, so q is the unique pre-image of p_0 in $SY(a_{\gamma_1}, \ldots, a_{\gamma_{n-1}})$ of the same rank and degree as p_0. Consequently $(Sj)q = p$, since Sj, being a homeomorphism, must preserve rank and degree. □

Lemma 6.1 is useful for generating indiscernibles over Y in $\mathcal{A}$. One simply begins with a suitable $p_0 \in SY$ and generates a Morley sequence in $\mathcal{A}$. A suitable p_0 is ranked, has many realizations in $\mathcal{A}$, and has a pre-image of the same rank and degree as p_0 over various extensions of Y in $\mathcal{A}$ (cf. [2], p. 232).

Lemma 6.2 (S. Shelah). Suppose $\mathcal{A}(b) \subset \mathcal{C} \in \mathcal{K}(T)$, the 1-type realized by b over $\mathcal{A}$ is ranked, and I is a set of indiscernibles over $\mathcal{A}$ in $\mathcal{C}$. Then there exists a finite $J \subset I$ such that $I-J$ is indiscernible over $\mathcal{A}(b)(J)$ in $\mathcal{C}$.

Proof. $\mathcal{A}(I)$ is the direct limit of the finitely generated extensions of $\mathcal{A}$ in $\mathcal{A}(I)$. Hence by 5.1 there exists a finite $J \subset I$ such that the rank and degree of the 1-type p_J realized by b over $\mathcal{A}(J)$ is the same as that of the 1-type p_I realized by b over $\mathcal{A}(I)$.

To verify the indiscernibility of $I-J$ over $\mathcal{A}(b)(J)$, let f be a one-one map of I_1 onto I_2, where I_1 and $I_2 \subset I-J$. f extends to an isomorphism

$$f : \mathcal{A}(J)(I_1) \overset{\approx}{\to} \mathcal{A}(J)(I_2)$$

which is the identity on $\mathcal{A}(J)$, since I is indiscernible over $\mathcal{A}$. Let p_i be the type of b over $\mathcal{A}(J)(I_i)$ $(i = 1,2)$. It suffices to show $Sfp_2 = p_1$.

The rank and degree rules imply p_1 and p_2 have the same rank and degree as p_J. Sfp_2 has the same rank and degree as p_2, since Sf is a homeomorphism. But then $Sfp_2 = p_1$ by the degree rule. □

Today's lecture concludes with a proof of Lemma 6.4, which requires some facts about prime model extensions summed up in Lemma 6.3 and proved tomorrow, facts that have little to do with rank.

Lemma 6.3. Suppose T is quasi-totally transcendental and $\mathcal{A} \in \mathcal{K}(T)$. Then there exists a prime model extension of $\mathcal{A}$. Every such extension

$\mathcal{B}$ is prime over every finitely generated extension of $\mathcal{A}$ in $\mathcal{B}$, and every element of it realizes a 1-type of $S\mathcal{A}$ which is both isolated and ranked.

Lemma 6.4 (S. Shelah). Suppose T is quasi-totally transcendental, $\mathcal{A} \in \mathcal{K}(T)$, and $\mathcal{B}$ is a prime model extension of $\mathcal{A}$. Then every set of indiscernibles over $\mathcal{A}$ in $\mathcal{B}$ is countable.

Proof. Suppose I is an uncountable set of indiscernibles over $\mathcal{A}$ in $\mathcal{B}$. For each formula F(x) in the language of $T \cup D\mathcal{B}$, Lemma 6.2 implies either (a) $\mathcal{B} \models F(\underline{i})$ for all but finitely many $i \in I$ or (b) $\mathcal{B} \models \sim F(\underline{i})$ for all but finitely many $i \in I$. Let p^I be the set of all F(x) that satisfy (a). $p^I \in S\mathcal{B}$ and is said to be the "average" type of I in $\mathcal{B}$.

Let $\mathcal{D}$ be any finitely generated extension of $\mathcal{A}$ in $\mathcal{B}$. By 6.2 the projection of p^I in $S\mathcal{D}$, call it p^I_D, is realized in $\mathcal{B}$ by all but finitely many $i \in I$. Hence p^I_D is ranked by 6.3. It follows from 5.2 that p^I is ranked. Choose $\mathcal{C} \subset \mathcal{B}$, finitely generated over $\mathcal{A}$, so that p^I_C has the same rank and degree as p^I. Then for every $\mathcal{E}$ such that $\mathcal{C} \subset \mathcal{E} \subset \mathcal{B}$, p^I_E has the same rank α and the same degree m.

By 6.2 there is an uncountable $K \subset I$ such that K is indiscernible over $\mathcal{C}$ in $\mathcal{B}$. Sequences $\{p_n\}$ and $\{i_n\}$ are defined by recursion: $p_0 \in S\mathcal{C}$ and every $i \in K$ realizes p_0; $i_n \in K$ and realizes p_n; p_{n+1} is realized by all but finitely $i \in K$ over $\mathcal{C}(i_0, \ldots, i_n)$. Note that rank $p_n = \alpha$ and $\deg p_n = m$.

According to 6.3 $\mathcal{C}(i_n | n < \omega)$ has a prime model extension $\mathcal{B}^* \subset \mathcal{B}$. Since $\mathcal{B}$ is prime over $\mathcal{C}$ by 6.3, there is an elementary monomorphism f of $\mathcal{B}$ into $\mathcal{B}^*$ such that $f|C = 1_C$. Let K^* be f[K]. Then K^* is indiscernible over $\mathcal{C}$ in $\mathcal{B}^*$. Define p_n^* to be the type realized by all but finitely many $i \in K^*$ over $\mathcal{C}(i_0, \ldots, i_{n-1})$. Clearly $p_0 = p_0^*$. Assume $p_n = p_n^*$ to see $p_{n+1} = p_{n+1}^*$. The rank rule implies $\operatorname{rank} p_{n+1}^* \leq \alpha$, since p_{n+1}^* is a pre-image of p_n^*. Let p^{K^*} be the "average type of K^* in $\mathcal{B}$. It is safe to assume I has been chosen among all uncountable sets of indiscernibles over $\mathcal{A}$ in $\mathcal{B}$ so as to minimize first the rank and then the degree of p^I. Thus $\operatorname{rank} p^{K^*} \geq \alpha$. But $p_{n+1}^* = p^{K^*}_{\mathcal{C}(i_0, \ldots, i_n)}$, hence $\operatorname{rank} p_{n+1}^* \geq \alpha$. Now the degree rule implies $p_{n+1}^* = p_{n+1}$, since p_{n+1}^* and p_{n+1} are pre-images of p_n of the same rank as p_n, and since $\deg p_{n+1} = \deg p_n = m$.

Let $p_\omega = \bigcup \{p_n | n < \omega\} = \bigcup \{p_n^* | n < \omega\}$. The uncountability of K and 6.2 require some $i^* \in K^*$ to realize p_ω. By 6.4 p_ω is an isolated point of $S\mathcal{C}(i_n | n < \omega)$. Hence there is a formula $F(x) \in p_\omega$ such that for every $G(x) \in p_\omega$,

$$T \cup D(\mathcal{C}(i_n | n < \omega)) \vdash F(x) \rightarrow G(x).$$

Fix n so that $F(x) \in p_n$. Since $i^* \neq i_n$, the formula $x \neq \underline{i}_n$ belongs to p_ω. But i_n satisfies F(x), hence must satisfy $x \neq \underline{i}_n$, an absolute impossibility. □

Tomorrow I will make a weak use of rank to prove the existence of prime model extensions of substructures of models of quasi-totally transcendental theories, and a strong use to prove their uniqueness.

G. e b. g.

7. Existence and Uniqueness of Prime Extensions

B. g. The first half of today's lecture will be devoted to Morley's proof of the existence of prime model extensions of substructures of models of T, where T satisfies a density condition much weaker than quasi-total transcendentality. The second half will concentrate on Shelah's proof of the uniqueness of the prime model extension, a proof whose original version [5] assumes total transcendentality but with trifling changes extends to the quasi-totally transcendental case.

7.1. Proposition. Suppose $\mathfrak{A} \in \mathcal{K}(T)$. (i) If $\mathfrak{B} \subset \mathfrak{A}$ and $\mathfrak{A} \models T$, then the 1-types of $S\mathfrak{B}$ realized in $\mathfrak{A}$ are dense in $S\mathfrak{B}$. (ii) If the 1-types of $S\mathfrak{A}$ realized in $\mathfrak{A}$ are dense in $S\mathfrak{A}$, then $\mathfrak{A} \models T$.

<u>Proof.</u> (i) Suppose $\{p \mid F(x, \underline{b}) \in p\}$ is a nonempty open subset of $S\mathfrak{B}$. Then the completeness of $T \cup D\mathfrak{B}$ implies $\mathfrak{A} \models F(\underline{a}, \underline{b})$ for some $a \in A$.

(ii) Extend $\mathfrak{A}$ to $\mathcal{C}$, a model of T. An induction on the logical complexity of sentences shows $\mathfrak{A} \prec \mathcal{C}$. For example suppose $\mathcal{C} \models (Ex)G(x)$. Then $\{p \mid G(x) \in p\}$ is a nonempty neighborhood of $S\mathfrak{A}$, and so contains some p realized by some $a \in A$. But then $\mathfrak{A} \models G(\underline{a})$.

$\square$

7.2. Proposition. If T is quasi-totally transcendental, then the isolated points of $S\mathcal{A}$ are dense in $S\mathcal{A}$ for every $\mathcal{A} \in \mathcal{K}(T)$.

Proof. Let N be a nonempty neighborhood of $S\mathcal{A}$. Choose a ranked $p \in N$ of the least possible rank. There is a neighborhood M such that $\{p\} = D^{\alpha}S\mathcal{A} \cap M$, where $\alpha = \text{rank } p$. But then $\{p\} = N \cap M$. □

7.3. Theorem. The isolated points of $S\mathcal{A}$ are dense in $S\mathcal{A}$ for every $\mathcal{A} \in \mathcal{K}(T)$ if and only if every $\mathcal{A} \in \mathcal{K}(T)$ has a prime model extension.

Proof. Assume the density condition holds in order to prove the existence of prime extensions. (The converse follows from 7.1(i) and a reduction to countable $\mathcal{A}$'s; cf. [2], Lemma 21.2 and Exercise 32.12.)

A chain $\{\mathcal{A}_\delta\}$ and a sequence $\{p_\delta\}$ are defined by recursion. (i) $\mathcal{A}_0 = \mathcal{A}$. (ii) $\mathcal{A}_\lambda = \bigcup \{\mathcal{A}_\delta \mid \delta < \lambda\}$ when λ is a limit ordinal. (iii) If $\mathcal{A}_\delta \models T$, then $\mathcal{A}_{\delta+1} = \mathcal{A}_\delta$. (iv) If $\mathcal{A}_\delta$ is not a model of T, then by 7.1(i) there is an isolated $p \in S\mathcal{A}_\delta$ not realized in $\mathcal{A}_\delta$; let one such p be p_δ, and let $\mathcal{A}_{\delta+1}$ be $\mathcal{A}_\delta(a_\delta)$, where a_δ realizes p_δ.

Let $\mathcal{C}$ be any model of T that extends $\mathcal{A}$. By 7.1(i) the construction of the $\mathcal{A}_\delta$'s can be mimicked inside $\mathcal{C}$. Thus $\mathcal{C}$ contains a chain $\{\mathcal{C}_\delta\}$ isomorphic to the chain $\{\mathcal{A}_\delta\}$. Since $\mathcal{C}$ is a set, there must be a γ such that $\mathcal{C}_\gamma = \mathcal{C}_{\gamma+1}$. But then $\mathcal{A}_\gamma$ is a model of T, and is prime over $\mathcal{A}$ because $\mathcal{C}$ is an arbitrary model of T extending $\mathcal{A}$. □

Suppose $\mathcal{A} \subset \mathcal{B} \in \mathcal{H}(T)$ and $n > 0$. Define $S_n\mathcal{A}$, the space of n-types of $T \cup D\mathcal{A}$, in the same manner $S\mathcal{A}$ ($= S_1\mathcal{A}$) was defined in Lecture 4. $\mathcal{B}$ is an <u>atomic extension</u> of $\mathcal{A}$ if for each n and $\langle b_1, \ldots, b_n \rangle \in B^n$, the n-type of $S_n\mathcal{A}$ realized by $\langle b_1, \ldots, b_n \rangle$ is isolated.

7.4. Proposition. Suppose $\mathcal{C} \supset \mathcal{B} \supset \mathcal{A}$. (i) If $\mathcal{C}$ is atomic over $\mathcal{B}$, and $\mathcal{B}$ is atomic over $\mathcal{A}$, then $\mathcal{C}$ is atomic over $\mathcal{A}$. (ii) If $\mathcal{C}$ is atomic over $\mathcal{A}$, and $\mathcal{B}$ is finitely generated over $\mathcal{A}$, then $\mathcal{C}$ is atomic over $\mathcal{B}$.

Proof. (i) Suppose $c \in C$ realizes the unique member of $S\mathcal{B}$ containing $F(\underline{b}, x)$, where $b \in B$. Let b realize the unique member of $S\mathcal{A}$ isolated by $G(x)$. Then c realizes the element of $S\mathcal{A}$ isolated by $(Ey)[G(y) \& F(y, x)]$.

(ii) It is safe to assume $\mathcal{C} = \mathcal{B}(c)$ and $\mathcal{B} = \mathcal{A}(b)$. Let $H(y, x)$ isolate the 2-type of $S_2\mathcal{A}$ realized by $\langle b, c \rangle$, and let q be the 1-type realized by c over $\mathcal{B}$. Then $H(\underline{b}, x)$ isolates q in $S\mathcal{B}$. □

7.5. Lemma. Suppose the isolated points of $S\mathcal{A}$ are dense in $S\mathcal{A}$ for every $\mathcal{A} \in \mathcal{H}(T)$. Let $\mathcal{B}$ be a prime model extension of $\mathcal{A}$. Then $\mathcal{B}$ is atomic over $\mathcal{A}$.

Proof. Since $\mathcal{B}$ is prime over $\mathcal{A}$, it suffices to find an atomic model extension of $\mathcal{A}$. 7.4(i) and an induction on δ show that the model $\mathcal{A}_\gamma$ ($= \cup \{\mathcal{A}_\delta \mid \delta \leq \gamma\}$) constructed in the proof of 7.3 is atomic over $\mathcal{A}$. □

7.6. Lemma. Suppose $\mathcal{C} \supset \mathcal{B} \supset \mathcal{A}$. If $\mathcal{C}$ is a prime model extension of $\mathcal{A}$, and $\mathcal{B}$ is finitely generated over $\mathcal{A}$, then $\mathcal{C}$ is prime over $\mathcal{B}$.

Proof. It is safe to assume $\mathcal{C}$ is the model $\mathcal{A}_\gamma$ $(= \cup \{\mathcal{A}_\delta \mid \delta \leq \gamma\})$ constructed in the proof of 7.3. Recall that $\mathcal{A}_{\delta+1} = \mathcal{A}_\delta(a_\delta)$, where a_δ realizes an isolated point of $S\mathcal{A}_\delta$. Define: $\mathcal{C}_0^* = \mathcal{B}$, $\mathcal{C}_\lambda^* = \cup\{\mathcal{C}_\delta^* \mid \delta < \lambda\}$, and $\mathcal{C}_{\delta+1}^* = \mathcal{C}_\delta^*(a_\delta)$. Thus $\mathcal{C} = \cup\{\mathcal{C}_\delta^* \mid \delta \leq \gamma\}$. 7.4 implies a_δ realizes an isolated point of $S\mathcal{C}_\delta^*$ for all $\delta < \gamma$. But then the argument given in 7.3 that $\mathcal{A}_\gamma$ is prime over $\mathcal{A}_0$ also shows $\mathcal{C}$ is prime over $\mathcal{C}_0^*$. □

Suppose $\mathcal{A} \subset \mathcal{B} \subset \mathcal{C} \in \mathcal{K}(T)$. $\mathcal{B}$ is <u>normal</u> over $\mathcal{A}$ in $\mathcal{C}$ if for all $p \in S\mathcal{A}$, all or none of the realizations of p in $\mathcal{C}$ belong to $\mathcal{B}$.

7.7. Lemma (L. Harrington). Suppose the isolated points of $S\mathcal{A}$ are dense in $S\mathcal{A}$ for every $\mathcal{A} \in \mathcal{K}(T)$. If $\mathcal{C}$ is an atomic model extension of $\mathcal{A}$, and $\mathcal{B}$ is normal over $\mathcal{A}$ in $\mathcal{C}$, then $\mathcal{C}$ is atomic over $\mathcal{B}$.

The truth of 7.7 is fairly evident when $\mathcal{A}$ is countable; thus most of the work of proving it consists of a reduction to the countable case via a downward Skolem-Lowenheim construction. (The basic downward construction shows every structure has a countable elementary substructure.) For details see [2], p. 206.

No further lemmas are needed for the proof of Shelah's uniqueness theorem, but one definition remains. Suppose $\mathcal{A} \in \mathcal{K}(T)$ and F is a closed subset of $S\mathcal{A}$. F is said to be ranked if every $p \in F$ is ranked.

The ordinal rank of a ranked F is the least $\gamma \geq \text{rank}\, p$ for all $p \in F$. If the ordinal rank of $F = \alpha$, then the compactness of $S\mathcal{A}$ implies at least one, but not infinitely many, $p \in F$ are of rank α. Let n be the highest degree attained among those p of rank α, and let d be the number of p's in F of rank α and degree n. The rank of a ranked F is (α, n, d). Let F(x) be a formula in the language of $T \cup D\mathcal{A}$. F(x) is <u>ranked over A</u> if $\{p \mid F(x) \in p\}$ is a ranked subset of $S\mathcal{A}$. The <u>rank</u> of <u>F(x) over A</u> is the rank of $\{p \mid F(x) \in p\}$. (The ranks are wellordered by "first differences".)

7.8. Theorem (S. Shelah). Suppose T is quasi-totally transcendental, $\mathcal{A}$ (respectively $\mathcal{A}^*$) $\in \mathcal{K}(T)$, $\mathcal{B}$ (respectively $\mathcal{B}^*$) is an atomic model extension of $\mathcal{A}$ (respectively $\mathcal{A}^*$), and $f : A \to A^*$ is an onto map such that

$$\langle \mathcal{B}, a \rangle_{a \in A} \equiv \langle \mathcal{B}^*, fa \rangle_{a \in A}.$$

Suppose further that every set of indiscernibles over $\mathcal{A}$ (respectively $\mathcal{A}^*$) in $\mathcal{B}$ (respectively $\mathcal{B}^*$) is countable. Let F(x) be a ranked formula over A. Define

$$C = A \cup \{b \mid b \in B \,\&\, \mathcal{B} \models F(\underline{b})\}$$
$$C^* = A^* \cup \{b \mid b \in B^* \,\&\, \mathcal{B}^* \models F(\underline{b})\}.$$

Then f can be extended to an onto map $g : C \to C^*$ such that $\langle \mathcal{B}, c \rangle_{c \in C} \equiv \langle \mathcal{B}^*, gc \rangle_{c \in C}$.

Proof. By induction on the rank (α, n, d) of F(x) over A. Choose

$p \in S\mathcal{A}$ such that $F(x) \in p$, rank $p = \alpha$ and $\deg p = n$. First suppose $d > 1$. Let $G(x) \in p$ isolate p in $D^{\alpha} S\mathcal{A}$. Define

$$D = A \cup \{b \mid b \in B \,\&\, \mathcal{B} \models F(\underline{b}) \,\&\, \sim G(\underline{b})\}$$

$$D^{*} = A \cup \{b \mid b \in B^{*} \,\&\, \mathcal{B}^{*} \models F(\underline{b}) \,\&\, \sim G(\underline{b})\}.$$

Then f can be extended to an onto map $h : D \to D^{*}$ such that $\langle \mathcal{B}, d \rangle_{d \in D} \equiv \langle \mathcal{B}, fd \rangle_{d \in D}$, since the rank of $F(x) \,\&\, \sim G(x)$ over A is $(\alpha, n, d-1)$. And h can be extended to the desired onto map $g : C \to C^{*}$, because the rank of $F(x) \,\&\, G(x)$ over D is at most $(\alpha, n, 1)$.

Now suppose $d = 1$. Thus p is the unique $q \in S\mathcal{A}$ such that $F(x) \in q$, rank $q = \alpha$ and $\deg q = n$. A "transcendence base" $\{b_{\delta} \mid \delta < \gamma_0\}$ for the realizations of p in $C-A$ is defined by recursion: (1) $p_0 = p$. (2) Choose $b_{\delta} \in C-A$ to realize p_{δ}; if no such b_{δ} exists, then $\gamma_0 = \delta$. (3) Choose $p_{\delta+1} \in S\mathcal{A}(b_{\gamma} \mid \gamma \leq \delta)$ to be a pre-image of p_{δ} of the same rank and degree as p_{δ}; if no such $p_{\delta+1}$ exists, then $\gamma_0 = \delta+1$. (4) $p_{\lambda} = \bigcup \{p_{\delta} \mid \delta < \lambda\}$.

If γ_0 is infinite, then $\{b_{\delta} \mid \delta < \gamma_0\}$ is a set of indiscernibles over $\mathcal{A}$ by 6.1, hence countable. Reorder the b_{δ}'s so that $\{b_{\delta} \mid \delta < \gamma_0\}$ becomes $\{b_m \mid m < \alpha_0\}$ for some $\alpha_0 \leq \omega$. Note that each $b \in C$ realizes a 1-type over $\mathcal{A}(b_m \mid m < \alpha_0)$ of rank $< \alpha$, or of rank α and degree $< n$.

Let $p^{*} = (Sf)^{-1} p \in S\mathcal{A}^{*}$. Repeat the above in C^{*} to obtain $\{b_m^{*} \mid m < \alpha_0^{*}\}$, and observe that $\alpha_0 = \alpha_0^{*}$ because $\mathcal{B}(\mathcal{B}^{*})$ is atomic over $\mathcal{A}(\mathcal{A}^{*})$ and $\langle \mathcal{B}, a \rangle_{a \in A} \equiv \langle \mathcal{B}^{*}, fa \rangle_{a \in A}$.

$g = \bigcup \{g_m \mid m < \omega\}$ is defined by recursion on m. Let $g_0 = f$, and fix $m \geq 0$. Assume domain $g_m \subset C$, range $g_m \subset C^*$, $\mathcal{B}(\mathcal{B}^*)$ is atomic over domain g_m (range g_m), and $\langle \mathcal{B}, c \rangle_{c \in \text{domain } g_m} \equiv \langle \mathcal{B}^*, g_m c \rangle_{c \in \text{domain } g_m}$. The definition of g_m has three cases. The first puts some b_k in the domain of g: The assumptions on g_m yield a $c^* \in C^*$ whose type over range g_m is the same as that of b_k over domain g_m. Let $g_{m+1} b_k = c^*$. Then 7.4(ii) implies the assumptions on g_m lift to g_{m+1}.

The second case puts some b_k^* in the range of g, and is similar to the first.

The purpose of the third case is to add to the domain of g (range of g) every $b \in C$ (C^*) such that b realizes some q over domain g_m (range g_m) of rank $< \alpha$, or of rank α and degree $< n$. Let $\{q_\gamma \mid \gamma < \tau\}$ be an enumeration of all such q's. Each q_γ is isolated in S(domain g_m) by some $F_\gamma(x)$. Happily the rank of $F_\gamma(x)$ over domain g_m is less than that of $F(x)$ over A.

Let q_γ^* be $(Sg_m)^{-1} q_\gamma$. Then $\{q_\gamma^* \mid \gamma < \tau\}$ is an enumeration of all $q \in S$ (range g_m) of rank $< \alpha$, or of rank α and degree $< n$, such that q is satisfied by some $b \in C^*$. Define $F_\gamma^*(x)$ similarly.

$g_{m+1} = \bigcup \{g_{m+1}^\gamma \mid \gamma < \tau\}$ is defined by recursion. Let $g_{m+1}^0 = g_m$, and fix $\gamma \geq 0$. Assume $\mathcal{B}$ ($\mathcal{B}^*$) is atomic over domain g_{m+1}^γ (range g_{m+1}^γ) and $\langle \mathcal{B}, b \rangle_{b \in \text{domain } g_{m+1}^\gamma} \equiv \langle \mathcal{B}^*, b \rangle_{b \in \text{domain } g_{m+1}^\gamma}$. The rank of $F_\gamma(x)$ over domain g_{m+1}^γ is less than that of $F(x)$ over A, so g_{m+1}^γ can be extended to $g_{m+1}^{\gamma+1}$ in such a way that the new domain is the old plus

$\{b \mid \mathcal{B} \models F_\gamma(\underline{b})\}$, the new range is the old plus $\{b \mid \mathcal{B}^* \models F_\gamma(\underline{b}^*)\}$, and the elementary equivalence of $\mathcal{B}$ and $\mathcal{B}^*$ over g^γ_{m+1} now holds over $g^{\gamma+1}_{m+1}$. By 7.7 $\mathcal{B}$ ($\mathcal{B}^*$) is atomic over domain $g^{\gamma+1}_{m+1}$ (range $g^{\gamma+1}_{m+1}$), since domain $g^{\gamma+1}_{m+1}$ (range $g^{\gamma+1}_{m+1}$) is normal over domain g^γ_{m+1} (range g^γ_{m+1}).

Let $g^\lambda_{m+1} = \cup \{g^\gamma_{m+1} \mid \gamma < \lambda\}$. Clearly domain g^λ_{m+1} is normal over domain g_{m+1}, so $\mathcal{B}$ is atomic over domain g^λ_{m+1}. □

Theorem 7.9. Suppose T is quasi-totally transcendental and $\mathcal{A} \in \mathcal{K}(T)$. Then any two prime model extensions of $\mathcal{A}$ are isomorphic over $\mathcal{A}$.

Proof. Suppose $\mathcal{B}$ and $\mathcal{C}$ are prime model extensions of $\mathcal{A}$. By 7.2 and 7.5 each $p \in S\mathcal{A}$ realized in $\mathcal{B}$ is an isolated point of $S\mathcal{A}$, and is in fact isolated by a formula ranked over A, since T is quasi-totally transcendental. Consequently 7.7 and 7.8 make it possible to build up an isomorphism between $\mathcal{B}$ and $\mathcal{C}$ by proceeding through normal extensions of $\mathcal{A}$ in $\mathcal{B}$ and $\mathcal{C}$. □

Tomorrow I will discuss the notion of differential closure for differentially closed fields of characteristic 0. G. e b. g.

8. Differentially Closed Fields

B. g. Today's lecture will be final but not complete. In the brief time left before the cessation of formalities, I will try to cover as much territory as possible.

The language of differential fields is that of fields together with a 1-place function symbol D. The theory of differential fields of characteristic 0 (DF_0) is the theory of fields of characteristic 0 with two additional axioms:

$$(x)(y)[D(x+y) = Dx + Dy],$$

$$(x)(y)[D(x \cdot y) = x \cdot Dy + y \cdot Dx].$$

DF_0 is a universal theory originated by Ritt [6], but DCF_0, the theory of differentially closed fields of characteristic 0, is a later invention of Robinson [7]. He proved somewhat indirectly that DF_0 has a model completion, necessarily unique by 3.1, and then defined DCF_0 to be the model completion of DF_0. Subsequently Blum [1] found the simple axioms for DCF_0 given below.

Let $\mathcal{A}$ be a differential field of characteristic 0. A <u>differential polynomial</u> over $\mathcal{A}$ in the variables x_i $(1 \leq i \leq m)$ is a polynomial over $\mathcal{A}$ in the variables $D^j x_i$ $(j \leq n_i)$. ($D^j x$ is the j-th derivative of x.) Let f(x) be a differential polynomial in one variable. Assume f(x) is in the following form:

$$(D^n x)^d + g_1(x)(D^n x)^{d-1} + \ldots + g_d(x),$$

where $g_i(x)$ $(1 \leq i \leq d)$ is a rational function over $\mathcal{A}$ in the variables $D^j x$ $(j < n)$. The order of f(x), ord f(x), is n, and the degree of f(x) is d. Let $\mathcal{B}$ be a differential field that extends $\mathcal{A}$. Suppose $b \in B-A$. b is a <u>generic</u> solution of f(x) over $\mathcal{A}$ if f(b) = 0 but $h(b) \neq 0$ for all h(x) over $\mathcal{A}$ of lower order than x. $c \in B-A$ is differential algebraic over $\mathcal{Q}$ if c is a solution of some differential polynomial over $\mathcal{Q}$; otherwise c is differential transcendental over $\mathcal{Q}$.

Blum's axioms for DCF_0 [1] are:

$$(Ex)[f(x) = 0 \;\&\; h(x) \neq 0] \quad (\text{ord}\, f(x) > \text{ord}\, g(x)),$$

$$(Ex)[f(x) = 0] \quad (\text{ord}\, f(x) = 0).$$

The next proposition, extracted from Seidenberg [8] and sketched in [2], p. 298, contains all the algebraic facts needed for the applications of model theory to DF_0 made by Robinson [7] and Blum [1].

Proposition 8.1. Let $\mathcal{Q}$ be a differential field of characteristic 0. (i) If b is differential algebraic over $\mathcal{Q}$, then there exists an f(x) such that b is a generic solution of f(x) over $\mathcal{Q}$, and such that all generic solutions of f(x) over $\mathcal{Q}$ are isomorphic over $\mathcal{Q}$. (ii) Each differential polynomial over $\mathcal{Q}$ has a generic solution in some extension of $\mathcal{Q}$. (iii) $\mathcal{Q}$ has a simple, differential transcendental extension, and all such extensions are isomorphic over $\mathcal{Q}$.

Theorem 8.2 (A. Robinson). DCF_0 is the model completion of DF_0.

Proof. For details of the proof of 8.2 see Blum [1] or [2]. Blum's approach is as follows. Suppose T and T^* are theories in the same language such that $T \subset T^*$, T is universal and every model of T can be extended to some model of T^*. Then T^* is the model completion of T if and only if every simple extension of every model $\mathcal{B}$ of T can be well approximated in every model $\mathcal{B}^*$ of T^* which extends $\mathcal{B}$. To say

$\mathcal{B}(c)$ is well approximated in $\mathcal{B}$ is to say for each quantifierless formula $F(x)$ in the language of $T \cup D\mathcal{B}$ such that $\mathcal{B}(c) \models F(c)$, there is a $b^* \in B^*$ such that $\mathcal{B}^* \models F(b^*)$.

8.1 (ii) implies each differential field $\mathcal{A}$ of characteristic 0 has a differentially closed extension, and 8.1(i) and (iii) imply every simple extension of $\mathcal{A}$ is well approximated in every differentially closed extension of $\mathcal{A}$. □

A theory T is ω-stable if $S\mathcal{A}$ is countable for every countable $\mathcal{A} \in \mathcal{K}(T)$, and is totally transcendental if every 1-type of $S\mathcal{A}$ is (Morley) ranked for every $\mathcal{A} \in \mathcal{K}(T)$.

Proposition 8.3 (M. Morley). T is ω-stable if and only if T is totally transcendental.

Proof. Suppose $p \in \mathcal{A} \in \mathcal{K}(T)$ is not ranked. 5.1 makes it possible to assume $\mathcal{A}$ and all other structures occurring in the following construction are countable. p must have distinct unranked pre-images p_0^1 and p_1^1 in $S\mathcal{A}_1$ for some $\mathcal{A}_1$ extending $\mathcal{A}_0$. p_0^1 and p_1^1 must "split" in a similar fashion in some $\mathcal{A}_2 \supset \mathcal{A}_1$. By continuing on to the limit, one obtains an $\mathcal{A}_\infty$ in which p has uncountably many pre-images, each of which corresponds to a path through a binary tree. (For the converse, cf. [2].) □

Theorem 8.3 (L. Blum). DCF_0 is totally transcendental.

Proof. An immediate consequence of 8.1 and 8.3. □

For each differential field $\mathcal{A}$ of characteristic 0, let $\overline{\mathcal{A}}$ be the unique prime differentially closed extension of $\mathcal{A}$ provided by 8.3 and 7.9. $\overline{\mathcal{A}}$ is called the <u>differential closure</u> of $\mathcal{A}$. Suppose $\overline{\mathcal{A}} \supset \mathcal{B} \supset \mathcal{A}$. Call f(x) irreducible over $\mathcal{A}$ if all generic solutions of f(x) over $\mathcal{A}$ are isomorphic over $\mathcal{A}$. $\mathcal{B}$ is a <u>normal</u> extension of $\mathcal{A}$ in $\overline{\mathcal{A}}$ if for each f(x) irreducible over $\mathcal{A}$, all or none of the generic solutions of f(x) in $\overline{\mathcal{A}}$ belong to $\mathcal{B}$. An extension $\mathcal{C}$ of $\mathcal{A}$ is said to be <u>atomic</u> over $\mathcal{A}$ if for each $n > 0$ and $\langle b_1, \ldots, b_n \rangle \in B^n$, there exists a finite system S of differential polynomial equalities and inequalities in n variables over $\mathcal{A}$ such that: $\langle b_1, \ldots, b_n \rangle$ is a solution of S, and all solutions of S are isomorphic over $\mathcal{A}$.

Summary Statement: DCF_0 admits elimination of quantifiers. Suppose $\mathcal{A}$ is a substructure of a model of DCF_0, i.e. $\mathcal{A}$ is a model of DF_0. Then among the models of DCF_0 extending $\mathcal{A}$ there is a unique prime one denoted by $\overline{\mathcal{A}}$. Every automorphism of every normal extension of $\mathcal{A}$ can be extended to an automorphism of $\overline{\mathcal{A}}$. $\overline{\mathcal{A}}$ is the unique model $\mathcal{B}$ of DCF_0 which extends $\mathcal{A}$, is atomic over $\mathcal{A}$, and has the property that every set of indiscernibles in $\mathcal{B}$ over $\mathcal{A}$ is countable.

The above summary statement applies to every theory T of algebraic type: simply replace DCF_0 by T, and DF_0 by the set W of universal sentences true in all models of T. (Since T is the model completion of some universal theory, it must also be the model completion of W.) Two further general properties of T not mentioned in the summary statement were discussed at the end of the third lecture for the case of algebraically closed fields. The second of the two implies there is an algorithm, first discovered by Seidenberg [8], for deciding whether or not a finite system S of differential polynomial equalities and inequalities in several variables over some differential field $\mathcal{A}$ has a solution in some extension of $\mathcal{A}$.

Let $\mathcal{Q}$ be the rational numbers construed as a differential field of characteristic 0 (D is necessarily 0 on $\mathcal{Q}$). L. Harrington [9] has shown $\overline{\mathcal{Q}}$ is computable. To be precise there is a one-one map of Q onto ω that transforms +, $\cdot$ and D into computable functions. Harrington's result is based on the finite basis theorem for differential ideals, and leaves open the following question. Is there an algorithm for deciding whether or not an arbitrary formula in the language of DCF_0 defines an isolated point of $S\mathcal{Q}$? The answer to the corresponding question for ACF_0 is yes, because there exists an effective method of factoring a polynomial into its irreducible factors.

Suppose $\mathcal{A} \in \mathcal{K}(T)$, $\mathcal{A} \subset \mathcal{B}$ and $\mathcal{B}$ is a model of T. $\mathcal{B}$ is said to be minimal over $\mathcal{A}$ if there is no $\mathcal{C} \subset \mathcal{B}$ such that $\mathcal{A} \subset \mathcal{C}$, $\mathcal{C}$ is a model of T and $\mathcal{C} \neq \mathcal{B}$. (For example the algebraic closure of a field is minimal over the field.) If T is quasi-totally transcendental, then 7.8 implies $\mathcal{B}$ is minimal over $\mathcal{A}$ if and only if $\mathcal{B}$ is atomic over $\mathcal{A}$ and every set of indiscernibles over $\mathcal{A}$ in $\mathcal{B}$ is finite. Kolchin [10], Rosenlicht [11] and Shelah [12] independently observed that $\overline{\mathcal{Q}}$ is not minimal over $\mathcal{Q}$. Kolchin shows that the equation $Dx = x^3 - x^2$ has an infinite set of algebraically independent, hence indiscernible, solutions in $\overline{\mathcal{Q}}$. His paper includes an illuminating exposition of the uniqueness of the differential closure from the viewpoint of differential algebra.

The definition of theory of algebraic type given in these lectures may be much too narrow. In particular it does not encompass C. Wood's theory of differentially closed fields of characteristic p (DCF_p) [13]. The language of the theory of perfect differential fields of characteristic p (PDF_p) is that of DF_0 augmented by a 1-place function symbol $\frac{1}{p}$ (p-th root). The axioms of PDF_p are those for fields of characteristic p, the two axioms for D of DF_0, and

$$(x)[Dx = 0 \rightarrow (x^{\frac{1}{p}})^p = x]$$

$$(x)[Dx \neq 0 \rightarrow x^{\frac{1}{p}} = 0].$$

Wood showed that PDF_p has a model completion, and defined DCF_p to be that model completion. (She also observed that the theory obtained by deleting the last two axioms of PDF_p has no model completion.) DCF_p is not quasi-totally transcendental, but it does satisfy the density condition of 7.3, and so every perfect differential field of characteristic p has a prime differentially closed extension. (The existence of the prime extension was obtained independently by Wood [14] and Shelah [12]. Wood's proof is simpler; it uses nothing more than the finite basis theorem for differential ideals.) DCF_p is stable [12], hence perfect differential fields of characteristic p do have unique prime differentially closed extensions by virtue of a result of Shelah [15] which holds for all stable theories that satisfy the density condition of 7. 3.

Perhaps quasi-total transcendentality is not the best choice for the final clause of the definition of theory of algebraic type. A less restrictive choice would be the density condition of 7. 3 coupled with quasi-stability. Call T quasi-stable if for each $f : \mathcal{A} \to \mathcal{B}$ in $\mathcal{H}(T)$ and each isolated $p \in S\mathcal{A}$, cardinality $(Sf)^{-1}p \leq (\text{cardinality } \mathcal{B})^{\omega}$. The theory of real closed fields is quasi-stable, but not stable. At this writing it is still possible that the density condition of 7. 3 suffices for uniqueness of prime model extensions, a doubtful consummation devoutly to be wished. Grazie dell'ascolto, e buon giorno.

References

[1] L. Blum, Generalized Algebraic Structures: A Model Theoretic Approach, Ph. D. Thesis, Massachusetts Institute of Technology, 1968.

[2] G. E. Sacks, Saturated Model Theory, Benjamin, Reading, Massachusetts, 1972.

[3] N. Jacobson, Basic Algebra I, Friedman, San Francisco, 1974.

[4] G. E. Sacks, Effective bounds on Morley rank, to appear.

[5] S. Shelah, Uniqueness and characterization of prime models over sets for totally transcendental first-order theories, Jour. Symb. Log., 37 (1972), 107-113.

[6] J. F. Ritt, Differential Algebra, Amer. Math. Soc. Colloquium Publ., 33 (1950).

[7] A. Robinson, On the concept of a differentially closed field, Bull. Res. Council Israel, Sect. F, 8 (1959), 113-128.

[8] A. Seidenberg, An elimination theory for differential algebra, Univ. California Publ. Math., 3 (1956), 31-65.

[9] L. Harrington, Recursively presentable prime models, Jour. of Symb. Log., 39 (1974), 305-309.

[10] E. R. Kolchin, Constrained extensions of differential fields, Advances in Mathematics, 12 (1974), 141-170.

[11] M. Rosenlicht, The nonminimality of the differential closure, Pacific Jour. of Math., to appear

[12] S. Shelah, Differentially closed fields, Israel Jour. of Math., 16 (1973), 314-328.

[13] C. Wood, The model theory of differential fields of characteristic $p \neq 0$, Proc. Amer. Math. Soc., 40 (1973), 577-584.

[14] C. Wood, Prime model extensions for differential fields of characteristic $p \neq 0$, Jour. of Symb. Log., 39 (1974), 469-477.

[15] S. Shelah, <u>Stability and the Number of Non-Isomorphic Models</u>, North-Holland, Amsterdam, in preparation.

Harvard University
Massachusetts Institute of Technology

Acknowledgement: The author is grateful to the Centro Internazionale Matematico Estivo for inviting him to give a course in their Summer Institute, and to the Institute for Advanced Study (Princeton) and the National Science Foundation for partially supporting him during the preparation of these lectures.

CENTRO INTERNAZIONALE MATEMATICO ESTIVO

(C.I.M.E.)

CONSTRUCTIONS IN MODEL THEORY

H. J. KEISLER

Corso tenuto a Bressanone dal 20 al 28 giugno 1975

CONSTRUCTIONS IN MODEL THEORY

H. Jerome Keisler

CONTENTS

CONSTRUCTIONS IN MODEL THEORY

H. Jerome Keisler
University of Wisconsin

In these lectures we survey several fundamental methods of constructing models. We examine which logics admit which constructions, and illustrate their use with examples. The lectures begin with methods available only in classical first order logic and proceed to methods available in the infinitary logics $\mathcal{L}_{\omega_1\omega}$ and $\mathcal{L}_{\infty\omega}$. We shall focus on three recent developments which have become more prominent in model theory since the publication of the book Chang and Keisler [7]. They are recursively saturated models (Lectures 2 and 3), model theoretic forcing (Lectures 5 and 6), and soft model theory (Lectures 4 and 8). Not all methods of constructing models are discussed here; for instance, ultraproducts and admissible sets have been left out. For background reading we suggest the books of Chang and Keisler [7], Keisler [13], and Morley [24]. This work was supported in part by NSF Grant number MPS74-06355A01.

Our notation is standard, but we mention here a few features which may need explanation. κ, λ always denote infinite cardinals, and $|X|$ is the cardinal of X. We sometimes use vector notation for finite sequences, $\vec{x} = \langle x_1, \ldots, x_n \rangle$. A language L is a set of relation, function, and constant symbols, and generates a first order predicate logic with equality. $\|L\|$ is the maximum of $|L|$ and ω. The notation $(\underline{A}, \vec{a}) \models \varphi$ means that in the model $\underline{A}$ for L, the n-tuple $\vec{a}$ satisfies the formula $\varphi(\vec{v})$. A theory T is a set of sentences of L. T is consistent if it has a model, and finitely satisfiable if every finite subset of T has a model. We use similar terminology for a set of formulas $\Gamma(v)$. Given a sublanguage $L_0 \subseteq L$ and a model $\underline{A}$ for L, $\underline{A} \upharpoonright L_0$ denotes the reduct of $\underline{A}$ to L_0. Given a subset B of the universe A of $\underline{A}$, $\underline{A} | B$ is the submodel of $\underline{A}$ with universe B. If U is a unary relation symbol of L interpreted by $U^{\underline{A}}$ in $\underline{A}$, the model $(\underline{A} \upharpoonright L_0) | U^{\underline{A}}$ is called a relativized reduct of $\underline{A}$ to L_0. The opposite of a reduct is an expansion. In case of a finite expansion $L_1 = L(P_1, \ldots, P_n)$ of L, we use the notation $\underline{A}_1 = (\underline{A}, P_1, \ldots, P_n)$ for an expansion of $\underline{A}$ to L_1.

H. J. Keisler

LECTURE 1. THE METHOD OF DIAGRAMS

The method of diagrams was invented by Henkin and A. Robinson around 1950. Both men used the method for a new proof of the Gödel completeness theorem and gave other applications. The idea of the method is to study models by adding new constant symbols to a language.

DEFINITION. Given a model $\underline{A}$ for L with universe A, the <u>diagram language</u> for $\underline{A}$ is the expansion L_A of L formed by adding a new constant symbol c_a to L for each $a \in A$. The <u>diagram expansion</u> of $\underline{A}$ is the model $\underline{A}_A$ for L_A such that each c_a is interpreted by a. More generally, given a subset $X \subseteq A$, the language L_X and model $\underline{A}_X$ for L_X are defined in the natural way. The <u>elementary diagram</u> of $\underline{A}$ is the complete theory $Th(\underline{A}_A)$ of the diagram expansion of $\underline{A}$.

In model theory the most useful consequence of the completeness theorem is the compactness theorem. As a warmup for these lectures we give a direct proof of the compactness theorem using diagrams.

COMPACTNESS THEOREM. Let T be a set of sentences of L. If every finite subset of T has a model, then T has a model.

Proof: Let $\|L\| = \kappa$. Form L_C by adding a set C of κ new constant symbols, and let φ_α, $\alpha < \kappa$ be an enumeration of the set of all sentences of L_C. Form an increasing chain T_α, $\alpha < \kappa$, of sets of sentences of L_C with the following properties:

(1) $T_0 = T$.

(2) Fewer than κ constant symbols from C occur in T_α.

(3) T_α is finitely satisfiable.

(4) If φ_α is consistent with T_α, then $\varphi_\alpha \in T_{\alpha+1}$.

(5) If φ_α is of the form $\exists x \psi(x)$ and is consistent with T_α, then $\psi(c) \in T_{\alpha+1}$ for some $c \in C$.

Let $T_\kappa = \bigcup_{\alpha<\kappa} T_\alpha$. Then T_κ contains T and is maximal finitely satisfiable in L_C. Let A be the set of all equivalence classes of constants $c \in C$ under the relation

$$c \sim d \quad \text{iff} \quad (c=d) \in T_\kappa .$$

There is a model $\underline{A}_C$ with universe A such that for each atomic sentence ψ of L_C.

(6) $\underline{A}_C \models \varphi$ if and only if $\varphi \in T_\kappa$.

By induction using (5) at the quantifier step, it can be shown that (6) holds for every sentence φ of L_C. Thus $\underline{A}_C$ is a model of T_κ, and the reduct $\underline{A}$ is a model of T.⊣

The compactness theorem has a great variety of applications. We give two examples concerning the important notion of a stable theory.

DEFINITION. Let $X \subseteq A$ and $a \in A$. The <u>type</u> of a over X is the set of all formulas $\varphi(v)$ in L_X satisfied by a in the model $\underline{A}_X$. A theory T is said to be κ-<u>stable</u> if for every model $\underline{A}$ and every set $X \subseteq A$ of power at most κ, $\underline{A}$ has at most κ different types over X.

THEOREM 1.1. (Morley 22). <u>Let</u> L <u>be countable. If a theory</u> T <u>in</u> L <u>is</u> ω-<u>stable</u>, <u>then</u> T <u>is</u> κ-<u>stable for all</u> κ.

<u>Proof</u>: Assume T is not κ-stable. Then there is a model $\underline{A}$ and an $X \subseteq A$ of power $\leq \kappa$ such that $\underline{A}$ has more than κ types over X. Call a formula $\varphi(v)$ of L_X <u>rich</u> over X if both $\varphi(v)$ and its negation belong to more than κ types over X. We may form a binary tree

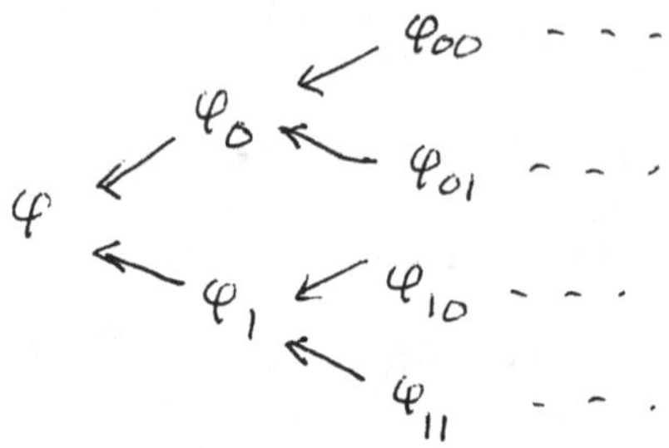

of rich formulas over X where the arrows represent logical implication and formulas on different branches are contradictory. Let Y be the set of constants in X which occur in this tree. Then Y is countable and the branches of the tree may be extended to 2^{ω} different consistent types over Y. By the compactness theorem there is a model $\underline{B}$ which has 2^{ω} types over Y, so T is not ω-stable. ⊣

THEOREM 1.2. (Morley 22). *Suppose* T *has a model* $\underline{A}$ *such that for some infinite* $X \subseteq A$ *and formula* $\varphi(u,v)$, *the relation* $\{\langle a,b\rangle : (\mathfrak{A},a,b) \models \varphi\}$ *linearly orders* X. *Then for all* κ, T *is not* κ-*stable*.

REMARK. If the variables u, v in $\varphi(u,v)$ are replaced by n-tuples, the converse also holds and gives a fundamental characterization due to Shelah 29 of theories unstable in all κ.

Proof: We use the fact that for each κ there is a linear ordering $\langle Y,<\rangle$ of power greater than κ with a dense subset Z of power κ . Let T' be the theory in L_Y whose axioms are T together with

$$\{ \varphi(x,y) \wedge \neg \varphi(y,x) : x < y \text{ in } Y \}.$$

T' is finitely satisfiable, hence has a model $\underline{A}_Y$. Since Z is dense in Y , if $x < y$ in Y - Z then there exists $z \in Z$ with $x < z < y$, whence $\varphi(x,z) \wedge \neg \varphi(y,z)$. Therefore x and y have different types over Z , and $\underline{A}$ has more than κ types over Z , so T is not κ-stable. ⊣

LECTURE 2. RECURSIVELY SATURATED MODELS.

Historically, saturated models have been a central tool in model theory, and gave a uniform method for proving a large number of results. However, saturated models cannot in general be shown to exist unless one assumes the continuum hypothesis or the existence of inaccessible cardinals, and most theories have no countable saturated models. In the next two lectures we discuss a pleasant recent discovery of Barwise and Schlipf 5, recursively saturated models. They have most of the advantages of saturated models but are easier to find.

For simplicity we assume that L is a finite language. We shall require only an intuitive knowledge of recursive sets. The main facts we need are that there are only countably many recursive sets, and whenever we write down a scheme to describe a set of formulas, the set is recursive.

DEFINITION. a model $\underline{A}$ for L is ω-<u>saturated</u> if for every finite set $X \subseteq A$, every set of formulas $\varphi(v)$ of L_X which is consistent with $Th(\underline{A}_X)$ is satisfiable in $\underline{A}_X$. $\underline{A}$ is said to be <u>recursively saturated</u> if for every finite set $X \subseteq A$, every recursive set of formulas $\varphi(v)$ which is consistent with $Th(\underline{A}_X)$ is satisfiable in $\underline{A}_X$.

Obviously, every ω-saturated model is recursively saturated. Every consistent theory has an ω-saturated model of power at most 2^{ω}, but not every consistent theory has a countable ω-saturated model. A counterexample is the theory of real closed ordered fields.

THEOREM 2.1. <u>Every consistent theory T has a finite or countable recursively saturated model.</u>

We leave the proof to the reader. The proof is like the proof of the compactness theorem, and depends on the fact that for each finite X, there are only countably many recursive sets of formulas in L_X .

We now develop some properties of recursively saturated models. It is easy to see that reducts and relativized reducts of recursively saturated models are recursively saturated.

DEFINITION. A model $\underline{A}$ is said to be ω-homogeneous if for any two n-tuples $\vec{a}$, $\vec{b} \in A^n$ such that $(\underline{A},a) \equiv (\underline{A},b)$, for all $c \in A$ there exists $d \in A$ such that

$$(\underline{A},\vec{a},c) \equiv (\underline{A},\vec{b},d) \quad .$$

THEOREM 2.2. (Schlipf 30). Every recursively saturated model is ω-homogeneous.

Proof: Let $X = \{\vec{a},\vec{b},c\}$. We must find $d \in A$ which satisfies the following set of formulas $\Sigma(v)$ in $\underline{A}_X$:

$$\{\varphi(\vec{a},c) \to \varphi(\vec{b},v) : \varphi(\vec{x},v) \text{ a formula of } L\} .$$

$\Sigma(v)$ is recursive, and is consistent with $Th(\underline{A}_X)$ because for each $\varphi(\vec{x},u)$, (1) $\Rightarrow$ (2) $\Rightarrow$ (3) .

(1) $\underline{A}_X \models \varphi(\vec{a},c)$

(2) $\underline{A}_X \models \exists u\, \varphi(\vec{a},u)$

(3) $\underline{A}_X \models \exists v\, \varphi(\vec{b},v)$.

Therefore $\Sigma(v)$ is satisfiable in $\underline{A}_X$. ⊣

DEFINITION. (Barwise). $\underline{A}$ is said to be resplendent if for every finite $X \subseteq A$ and every sentence φ in an expansion L'_X of L_X, if φ is consistent with $Th(\underline{A}_X)$ then some expansion $\underline{A}'_X$ of $\underline{A}_X$ satisfies φ.

This notion is analogous to recursive saturation but instead of a recursive set of formulas $\varphi(v)$ we have a single sentence φ with extra relation or function symbols. Using a classical result of Kleene in recursion theory, Barwise showed that every resplendent model is recursively saturated. In general, an uncountable recursively saturated model need not be resplendent. We shall now prove that every countable recursively saturated model is resplendent.

THEOREM 2.3. (Barwise). Every countable recursively saturated model $\underline{A}$ is resplendent.

Proof: We work in the language L'_A and use an argument similar to the compactness proof. Let $\theta_0, \theta_1, \ldots$ be a list of all sentences of L'_A. Let $X \subseteq A$ be finite and let φ be a sentence of L'_X consistent with $Th(\underline{A}_X)$. We form a sequence $\varphi_0, \varphi_1, \varphi_2, \cdots$ of sentences of L'_A with the following properties.

(1) $\models \varphi_{n+1} \rightarrow \varphi_n$

(2) $\models \psi_0 \to \varphi$

(3) ψ_n is consistent with $Th(\underline{A}_A)$.

(4) If θ_n is consistent with $Th(\underline{A}_A)$ and ψ_n , then $\models \psi_{n+1} \to \theta_n$.

(5) If θ_n is of the form $\exists v\theta(v)$ and is consistent with $Th(\underline{A}_A)$ and ψ_n , then $\models \psi_{n+1} \to \theta(a)$ for some $a \in A$.

Recursive saturation is needed for property (5). Given ψ_n, assume the hypotheses of (5). We have $\psi_n, \theta \in L'_Y$ for some finite $Y \subseteq A$. There is a recursive set $\Sigma(v)$ of formulas of L_Y which has exactly the same consequences as $\psi_n \wedge \theta(v)$ in L_Y . Then $\Sigma(v)$ is consistent with $Th(\underline{A}_Y)$. By recursive saturation there exists $a \in A$ such that a satisfies $\Sigma(v)$ in $\underline{A}_Y$. It follows that $\theta(a)$ is consistent with $Th(\underline{A}_A)$ and ψ_n , so we take $\psi_{n+1} = \psi_n \wedge \theta(a)$.

Now let $T = \{\psi_0, \psi_1, \psi_2, \ldots\}$. It follows from (1) - (5) that $\underline{A}_A$ has an expansion $\underline{A}'_A$ to L'_A which is a model of T . Therefore $\underline{A}'$ satisfies φ . $\dashv$

LECTURE 3. EXPANSIONS AND PARTIAL ISOMORPHISMS.

A simple but powerful method in model theory is the method of expansions. To prove a new theorem for a given language L , use an old theorem applied to an expanded language L' . We shall illustrate this method using recursively saturated models in an expanded language. For simplicity we assume that the original language L has predicate symbols only, and is finite, $L = \{P_1,\ldots,P_r\}$.

DEFINITION. Let $\underline{A}$ and $\underline{B}$ be two models for L with disjoint universes. The model pair $(\underline{A},\underline{B})$ is the model for the expanded language

$$L(U,V) = \{P_1,\ldots,P_r,U,V\}$$

given by

$$(\underline{A},\underline{B}) = \langle A \cup B, P_1^A \cup P_1^B , \ldots, P_r^A \cup P_r^B , A, B\rangle .$$

Thus both $\underline{A}$ and $\underline{B}$ are relativized reducts of $(\underline{A},\underline{B})$.

We shall use the fact that any model pair $(\underline{A},\underline{B})$ is elementarily equivalent to a countable recursively saturated model pair $(\underline{C},\underline{D})$. First we introduce the notion of a partial isomorphism, which is due to Karp [1] and is important in infinitary logic.

DEFINITION. A <u>partial isomorphism</u> $I:\underline{A} \cong_p \underline{B}$ is a relation I on the set of finite sequences $(\vec{a},\vec{b})$ of elements of $A \times B$ such that:

(i) $\emptyset \; I \; \emptyset$.

(ii) If $\vec{a} \; I \; \vec{b}$ then $(\underline{A},\vec{a})$ and $(\underline{B},\vec{b})$ satisfy the same atomic sentences.

(iii) If $\vec{a} \; I \; \vec{b}$ then for all $c \in A$ there exists $d \in B$ such that $(\vec{a},c) \; I \; (\vec{b},d)$, and vice versa.

$\underline{A} \cong \underline{B}$ means that $\underline{A}$ and $\underline{B}$ are isomorphic, and $\underline{A} \cong_p \underline{B}$ means that $\underline{A}$ and $\underline{B}$ are partially isomorphic. We shall see later that $\underline{A} \cong_p \underline{B}$ is stronger than $\underline{A} \equiv \underline{B}$ but weaker than $\underline{A} \cong \underline{B}$.

THEOREM 3.1. (Karp 11). <u>If $\underline{A}$ and $\underline{B}$ are countable and partially isomorphic, then they are isomorphic.</u>

The proof, which is a back and forth construction, is left to the reader.

THEOREM 3.2. (Schlipf 30). <u>If $\underline{A} \equiv \underline{B}$ and the model pair $(\underline{A},\underline{B})$ is recursively saturated, then $\underline{A} \cong_p \underline{B}$. Thus if $\underline{A}$ and $\underline{B}$ are also countable, then $\underline{A} \cong \underline{B}$.</u>

<u>Proof</u>: This is essentially a generalization of Theorem 2.2. We skip the details but remark that the partial

isomorphism I is given by $\vec{a}\, I\, \vec{b}$ if and only if $(\underline{A},\vec{a}) \equiv (\underline{B},\vec{b})$. ⊣

We shall now give some applications of model pairs of recursively saturated models. In each proof below, $(\underline{C},\underline{D})$ will denote a countable recursively saturated model elementarily equivalent to a given model pair $(\underline{A},\underline{B})$.

ROBINSON CONSISTENCY THEOREM. Suppose $\underline{A}$ and $\underline{B}$ are models for languages L_1 and L_2 such that $L_0 = L_1 \cap L_2$ and $\underline{A}\restriction L_0 \equiv \underline{B}\restriction L_0$. Then $Th(\underline{A}) \cup Th(\underline{B})$ is consistent in the language $L_1 \cup L_2$.

Proof. The notion of a model pair can be extended in the natural way to models for different languages. Form the model pair $(\underline{A},\underline{B})$ and let $(\underline{C},\underline{D})$ be an elementarily equivalent countable recursively saturated model pair. Then $\underline{C}\restriction L_0 \equiv \underline{D}\restriction L_0$, so by Theorem 3.2, $\underline{C}\restriction L_0 \cong \underline{D}\restriction L_0$. Make $\underline{D}$ into a model for $L_1 \cup L_2$ by interpreting the symbols of $L_2 - L_1$ by their images under the isomorphism. The resulting model is a model of $Th(\underline{A}) \cup Th(\underline{B})$. ⊣

EXAMPLE 1. (Tarski 31). The theory of real closed ordered fields is complete.

Proof. Let $\underline{A}$ and $\underline{B}$ be real closed ordered fields.

Let $(\underline{C},\underline{D})$ be an elementarily equivalent countable recursively saturated pair. It suffices to show that $\underline{C} \cong_p \underline{D}$. Using recursive saturation, it can be shown that the following relation I is a partial isomorphism.

$\vec{a}\ I\ \vec{b}$ if and only if for any polynomial $P(\vec{x})$ with rational coefficients, $P(\vec{a}) > 0$ iff $P(\vec{b}) > 0$. ⊣

EXAMPLE 2. (Presburger 25). The following axioms T characterize the complete theory of the integers with order and addition, i. e. the model $\langle Z,+,-,0,1,\leq\rangle$.

1. Abelian group axioms with +, -, 0 .
2. Axioms for linear order.
3. $x \leq y \rightarrow x + z \leq y + z$.
4. 1 is the least element greater than 0 .
5. For each positive integer n , the axiom

$$\forall x\ (\ n|x \vee n|x+1 \vee \ldots \vee n|x+(n-1)\ )$$

where $n|x$ means $\exists y(\underbrace{y + \ldots + y}_{n} = x)$.

Proof. Let $\underline{A}$, $\underline{B}$ be models of T, and form $(\underline{C},\underline{D})$ as before. Using recursive saturation, one can check that the relation I defined as follows is a partial isomorphism from $\underline{C}$ to $\underline{D}$.

$\vec{a}$ I $\vec{b}$ if and only if for each linear function $P(\vec{x})$ with integer coefficients, and each positive integer n ,

$$P(\vec{a}) > 0 \text{ if and only if } P(\vec{b}) > 0 ,$$

$$n \mid P(\vec{a}) \text{ if and only if } n \mid P(\vec{b}) . \quad \dashv$$

LYNDON HOMOMORPHISM THEOREM. (Lyndon 20). *Suppose every positive sentence true in* $\underline{A}$ *is true in* $\underline{B}$. *Then there exist models* $\underline{C} \equiv \underline{A}$ *and* $\underline{D} \equiv \underline{B}$ *such that* $\underline{D}$ *is a homomorphic image of* $\underline{C}$.

In the above theorem, a *positive sentence* is a sentence whose only connectives are $\wedge$ and $\vee$. A *homomorphism* is a mapping f of C into D such that every atomic formula satisfied by $\vec{a}$ in $\underline{C}$ is satisfied by $f\vec{a}$ in $\underline{D}$. $\underline{D}$ is a *homomorphic image* of $\underline{C}$ if there is a homomorphism of $\underline{C}$ onto $\underline{D}$.

Proof. Let $(\underline{C},\underline{D})$ be countable, recursively saturated, and elementarily equivalent to $(\underline{A},\underline{B})$. We show that $\underline{D}$ is a homomorphic image of $\underline{C}$. Let $\vec{a}$ I $\vec{b}$ mean that every positive sentence true in $(\underline{C},\vec{a})$ is true in $(\underline{D},\vec{b})$. Using recursive saturation, I is a partial homomorphism in the natural sense. By a back and forth argument there is a homomorphism of $\underline{C}$ onto $\underline{D}$. $\dashv$

LECTURE 4. SOFT MODEL THEORY AND LINDSTROM'S THEOREM

Lindstrom's theorem shows, roughly, that classical first order logic is the strongest logic which has the properties we used in the first three lectures, namely the compactness theorem, the Löwenheim-Skolem theorems, and the method of expansions. In order to state such a theorem, we need a general notion of a logic. Lindstrom's theorem is the beginning of a subject called soft model theory, which studies logics from an abstract viewpoint. It has been developed recently by Barwise, Feferman, Friedman, Makowsky, Shelah, Stavi, and others.

DEFINITION. An abstract logic is a pair of classes $(\mathcal{L}^*, \models^*)$ with the following properties. $\mathcal{L}^*$ is the class of sentences and $\models^*$ is the satisfaction relation.

(i) Occurrence property. For each $\varphi \in \mathcal{L}^*$ there is associated a finite language L_φ, called the set of symbols occurring in φ. $\underline{A} \models^* \varphi$ is a relation between sentences φ and models $\underline{A}$ for languages $L \supseteq L_\varphi$.

(ii) Expansion property. If $\underline{A} \models^* \varphi$ and $\underline{B}$ is an expansion of $\underline{A}$ to a larger language, then $\underline{B} \models^* \varphi$.

(iii) <u>Isomorphism property</u>. If $\underline{A} \cong \underline{B}$ and $\underline{A} \models^* \varphi$ then $\underline{B} \models^* \varphi$.

(iv) <u>Closure property</u>. $\mathcal{L}^*$ contains all atomic sentences, is closed under the connectives $\wedge$, $\vee$, $\neg$, and the usual rules of satisfaction for atomic formulas and connectives hold for $\models^*$.

(v) <u>Quantifier property</u>. For each constant symbol $c \in L_\varphi$ there are sentences $\forall v_c \varphi$ and $\exists v_c \varphi$ with the set of symbols $L_\varphi - \{c\}$ such that

$\underline{A} \models^* \forall v_c \varphi$ iff for all $a \in A$, $(\underline{A}, a) \models^* \varphi$,

$\underline{A} \models^* \exists v_c \varphi$ iff for some $a \in A$, $(\underline{A}, a) \models^* \varphi$.

(vi) <u>Relativization property</u>. For each sentence φ in $\mathcal{L}^*$ and relation $R(v,\vec{b})$ with $R, \vec{b} \notin L_\varphi$, there is a new sentence $\varphi^{R(v,\vec{b})}$, read φ <u>relativized</u> to $R(v,\vec{b})$, such that whenever $\underline{B} = \underline{A} | \{a \in A : R(a,\vec{b})\}$ is a submodel of $\underline{A}$,

$$(\underline{A}, R, \vec{b}) \models^* \varphi^{R(v,\vec{b})} \quad \text{iff} \quad \underline{B} \models^* \varphi .$$

Before stating any theorems we list some of the most widely studied examples of abstract logics. Each example is described by adding one or more rules for building formulas to the rules for classical first order logic, and taking for $\mathcal{L}^*$ the class of formulas which are sentences (i.e. have no free variables) and have only finitely many nonlogical symbols.

EXAMPLE 1. $\mathcal{L}_{\omega\omega}$. This is classical first order logic.

EXAMPLE 2. $\mathcal{L}_{\omega_1\omega}$. Add the rule: if Φ is a countable set of formulas, then $\bigwedge\Phi$ and $\bigvee\Phi$ are formulas.

EXAMPLE 3. $\mathcal{L}_{\infty\omega}$. Add the rule: if Φ is a set of formulas, then $\bigwedge\Phi$ and $\bigvee\Phi$ are formulas.

EXAMPLE 3. $\mathcal{L}_{\omega_1\omega_1}$. Add the rules: if Φ is a countable set of formulas, then $\bigwedge\Phi$ and $\bigvee\Phi$ are formulas. If φ is a formula and U is a countable set of variables, then $\forall U\varphi$ and $\exists U\varphi$ are formulas.

EXAMPLE 5. $\mathcal{L}(Q_\kappa)$. Add the rule: if φ is a formula, $(Q_\kappa v)\varphi$ is a formula. $\underline{A} \models^* (Q_\kappa v)\varphi$ iff there are at least κ elements $a \in A$ such that $(\underline{A}, a) \models^* \varphi$.

EXAMPLE 6. $\mathcal{L}_{WO}$. Add the rule: if φ is a formula, then $(Q_{WO}uv)\varphi$ is a formula. $\underline{A} \models^* (Q_{WO}uv)\varphi$ iff the relation $\{\langle a,b\rangle : (\underline{A},a,b) \models^* \varphi$ is a well ordering$\}$.

We next introduce several properties of abstract logics which are motivated by basic results for $\mathcal{L}_{\omega\omega}$.

DEFINITION. An abstract logic $(\mathcal{L}^*, \models^*)$ is said to be <u>countably compact</u> if for every countable set $T \subseteq \mathcal{L}^*$, if every finite subset of T has a model then T has a model.

The <u>Löwenheim number</u> of $(\mathcal{L}^*, \models^*)$ is the least cardinal κ such that every sentence of $\mathcal{L}^*$ which has a model has a model of power at most κ.

The <u>Hanf number</u> of $(\mathcal{L}^*, \models^*)$ is the least cardinal λ such that every sentence of $\mathcal{L}^*$ which has a model of power at least λ has models of arbitrarily large power.

$(\mathcal{L}^*, \models^*)$ has the <u>Karp property</u> if any two models which are partially isomorphic are elementarily equivalent with respect to $(\mathcal{L}^*, \models^*)$; in symbols, $\underline{A} \cong_p \underline{B}$ implies $\underline{A} \equiv_{\mathcal{L}^*} \underline{B}$.

The classical downward and upward Löwenheim-Skolem theorems show that $\mathcal{L}_{\omega\omega}$ has Löwenheim number ω and Hanf number ω. The Löwenheim number and Hanf number need not exist if $\mathcal{L}^*$ is a proper class, for example $\mathcal{L}_{\infty\omega}$.

<u>Note</u>: If L has function or constant symbols, then $I: \underline{A} \cong_p \underline{B}$ is defined to mean that the corresponding purely relational models are partially isomorphic by I.

The following table summarizes the behavior of our example

Logic	countably compact	Löwenheim number	Hanf number	Karp prop.
$\mathcal{L}_{\omega\omega}$	yes	ω	ω	yes
$\mathcal{L}_{\omega_1\omega}$	no	ω	$\beth_{\omega_1}$	yes
$\mathcal{L}_{\infty\omega}$	no	none	none	yes
$\mathcal{L}_{\omega_1\omega_1}$	no	2^{ω}	very large	no
$\mathcal{L}(Q_\omega)$	no	ω	$\beth_{\omega_1^{ck}}$	yes
$\mathcal{L}(Q_{\omega_1})$	yes	ω_1	$\beth_\omega$	no
$\mathcal{L}_{WO}$	no	ω	very large	yes

THEOREM 4.1. (Hanf 9). <u>If $\mathcal{L}^*$ is a set, then the Löwenheim number and Hanf number of $(\mathcal{L}^*, \models^*)$ exist.</u>

<u>Proof.</u> For each consistent sentence $\varphi \in \mathcal{L}^*$, let $\kappa(\varphi)$ be the least cardinal of a model of φ. Let $\lambda(\varphi) = \omega$ if φ has arbitrarily large models, and $\lambda(\varphi)$ be the supremum of the cardinals of models of φ otherwise. Then $(\mathcal{L}^*, \models^*)$ has Löwenheim number and Hanf number

$$\kappa = \sup\{\kappa(\varphi) : \varphi \in \mathcal{L}^*\}, \quad \lambda = \sup\{\lambda(\varphi) : \varphi \in \mathcal{L}^*\}. \quad \dashv$$

THEOREM 4.2. (Barwise 3). <u>If $(\mathcal{L}^*, \models^*)$ has Löwenheim number ω, then it has the Karp property.</u>

<u>Proof.</u> Suppose $I : \underline{A} \cong_p \underline{B}$ but $\underline{A} \models^* \varphi$, $\underline{B} \models^* \neg\varphi$. Let A' be the set of finite sequences of elements of A and let $F : A' \times A \to A'$ be the function

$$F(\langle a_1, \ldots, a_m \rangle, b) = \langle a_1, \ldots, a_m, b \rangle.$$

Define B' and G analogously. Since $\underline{A} \cong_p \underline{B}$ we can make $A \cap B$ be the set of constants and form the expanded model $(\underline{A},\underline{B},A',B',F,G,I)$. By the closure, quantifier, and relativization properties there is a sentence $\varphi \in \mathcal{L}^*$ which states that

$$\underline{A} \models^* \varphi\ ,\quad \underline{B} \models^* \neg\varphi,\quad I:\underline{A} \cong_p \underline{B}\ .$$

Since the Löwenheim number is ω, φ has a countable model $(\underline{A}_0,\underline{B}_0,A_0',B_0',F_0,G_0,I_0)$. $\underline{A}_0$ and $\underline{B}_0$ are countable and partially isomorphic, hence isomorphic. But $\underline{A}_0 \models^* \varphi$, and $\underline{B}_0 \models^* \neg\varphi$, contradicting the isomorphism property. We conclude that the Karp property holds. ⊣

LINDSTROM'S THEOREM. (Lindstrom [7]) $\mathcal{L}_{\omega\omega}$ *is the only abstract logic which has Löwenheim number* ω *and one of the following:*

(i) *Countable compactness;*

(ii) *Hanf number* ω .

Proof. We prove the case (i). Let $(\mathcal{L}^*, \models^*)$ be an abstract logic with Löwenheim number ω and countable compactness. We must show that every $\varphi \in \mathcal{L}^*$ is equivalent to some $\psi \in \mathcal{L}_{\omega\omega}$, i.e. for all $\underline{A}$,

$$\underline{A} \models^* \varphi \quad \text{iff} \quad \underline{A} \models \psi\ .$$

It is sufficient to consider a finite language L with only relation symbols. Given two models $\underline{A}$ and $\underline{B}$ for L and m-tuples $\vec{a}$ and $\vec{b}$, we define $\vec{a}\, I_n\, \vec{b}$ inductively as follows. $\vec{a}\, I_0\, \vec{b}$ means that $\vec{a}$ and $\vec{b}$ satisfy the same atomic formulas. $\vec{a} I_{n+1} \vec{b}$ holds if and only if:

(1) for all $c \in A$ there exists $d \in B$ with $\vec{a},c\ I_n\ \vec{b},d$,

(2) vice versa.

$\underline{A} \equiv_n \underline{B}$ means $\emptyset\, I_n\, \emptyset$. There is a finite set Ψ_n of sentences of $\mathcal{L}_{\omega\omega}$ such that $\underline{A} \equiv_n \underline{B}$ if and only if $\underline{A}$ and $\underline{B}$ satisfy the same sentences of Ψ_n.

Let $\varphi \in \mathcal{L}^*$, $L_\varphi \subseteq L$. It suffices to show that for some n,

(3) $\underline{A} \equiv_n \underline{B}$ and $\underline{A} \models^* \varphi$ implies $\underline{B} \models^* \varphi$,

for then φ must be equivalent to a Boolean combination of Ψ_n. Suppose (3) fails for all n. Choose models $\underline{A}_n, \underline{B}_n \ni$

(4) $\underline{A}_n \equiv_n \underline{B}_n$, $\underline{A}_n \models^* \varphi$, and $\underline{B}_n \models^* \neg\varphi$.

We can now form an expanded model

$$(\underline{C}, \underline{D}, R, S, \omega, \leq, \ldots)$$

such that for each n,

$$\underline{A}_n = \underline{C} \mid \{a : R(a,n)\}, \quad \underline{B}_n = \underline{D} \mid \{b : S(b,n)\},$$

and there is a sentence $\theta \in \mathcal{L}^*$ stating that (4) holds for all n. By countable compactness, θ has a model

$$(\underline{C}', \underline{D}', R', S', \omega', \leq', \ldots)$$

such that $\langle \omega', \leq' \rangle$ has a nonstandard element H. Then

$$\underline{A}'_H \models^* \varphi , \quad \underline{B}'_H \models^* \neg \varphi , \quad \text{and} \quad \underline{A}'_H \equiv_H \underline{B}'_H .$$

It follows that the relation between m-tuples given by

$$\vec{a} \; J \; \vec{b} \quad \text{iff} \quad \vec{a} \; I_{H-m} \; \vec{b}$$

is a partial isomorphism between $\underline{A}'_H$ and $\underline{B}'_H$. But $(\models^*, \mathcal{L}^*)$ has the Karp property by Theorem 4.2, so we have a contradiction. We conclude that φ is equivalent to some sentence in $\mathcal{L}_{\omega\omega}$. $\dashv$

LECTURE 5. BASIC MODEL THEORY AND ROBINSON FORCING.

In this section we present an alternative development of model theory which is more appropriate for the study of algebra. We concentrate on basic formulas, i. e. atomic formulas and their negations. Much of model theory can be developed in either of two parallel ways, emphasizing basic or arbitrary formulas. Basic model theory seems more special, but it is often more general because one can make every formula equivalent to a basic formula as follows. Add a new relation symbol $R(\vec{v})$ to the language L for each formula $\varphi(\vec{v})$, expand each theory of L by adding the axioms $\forall \vec{v}\, R(\vec{v}) \leftrightarrow \varphi(\vec{v})$, and uniquely expand each model for L to satisfy these axioms.

The basic analogue of the elementary diagram of $\underline{A}$ is the ordinary diagram of $\underline{A}$, denoted by $D(\underline{A})$.$D(\underline{A})$ is defined as the set of all basic sentences true in $\underline{A}_A$. In basic model theory we concentrate on inductive theories rather than arbitrary theories. A theory T is said to be inductive if it has a set of $\forall\exists$ axioms. It is easy to see that the union of any increasing chain of models of an inductive theory

T is again a model to T . The notion of a κ-stable theory has the following basic analogue.

DEFINITION. Given a set $X \subseteq A$ and an element $a \in \mathbf{A}$, the _basic type of_ a _over_ X is the set of all basic formulas $\varphi(v)$ in L_X satisfied by a in $\underline{A}_X$. A theory T is said to by _basically_ κ_-stable_ if for every model $\underline{A}$ of T and set $X \subseteq A$ of power at most κ , $\underline{A}$ has at most κ basic types over X .

The analogues of Theorems 1.1 and 1.2 hold for basic stability and have similar proofs. Thus every basically ω-stable theory is basically κ-stable for all κ . Basic stability is a more natural notion in algebra and it is easier to find examples.

Examples of basically ω-stable theories: Fields, abelian groups, integral domains, differential fields of characteristic zero.

Examples of theories which are basically stable in some but not all κ (in fact for exactly those κ with $\kappa = \kappa^{\omega}$): Differential fields of characteristic p with a pth root symbol, R-modules where R is a fixed sufficiently non-trivial countable ring.

Examples of theories which are basically unstable in all cardinals κ : Linear order, ordered fields, fields with valuation, groups, division rings, boolean algebras.

We shall now develop Robinson's notion of finite forcing. In the next lecture we generalize forcing to the infinitary logic $\mathcal{L}_{\omega_1\omega}$. We shall see then that forcing is the key constuction which gives $\mathcal{L}_{\omega_1\omega}$ a well behaved model theory.

Let L be a countable language with an infinite set C of constant symbols (there may be other constant symbols outside C). Hereafter we let T be a consistent inductive theory in which no $c \in C$ occurs.

DEFINITION. A <u>condition</u> for T is a finite set of basic sentences which is consistent with T . p, q, ... denote conditions for T .

DEFINITION. The relation p <u>forces</u> φ , denoted by $p \Vdash \varphi$ or if necessary $p \Vdash_T \varphi$, is between conditions p and arbitrary sentences φ of L . It is defined by:

If φ is atomic, $p \Vdash \varphi$ iff $\varphi \in p$.

$p \Vdash \neg\varphi$ iff there is no condition $q \supseteq p$ with $q \Vdash \varphi$.

$p \Vdash \exists u \varphi(u)$ iff $p \Vdash \varphi(c)$ for some $c \in C$.

$p \Vdash \varphi \vee \psi$ iff $p \Vdash \varphi$ or $p \Vdash \psi$.

In this definition we take $\neg$, $\vee$,and $\exists$ as fundamental.

We say that p <u>weakly forces</u> φ , $p \models^{w} \varphi$, if $p \Vdash \neg\neg\varphi$.

LEMMA 5.1. (i) <u>If</u> $p \subseteq q$ <u>and</u> $p \Vdash \varphi$, <u>then</u> $q \Vdash \varphi$.

(ii) <u>If</u> $\varphi \in p$ <u>then</u> $p \Vdash \varphi$.

(iii) $p \Vdash^{w} \varphi$ <u>if and only if for all</u> $q \supseteq p$ <u>there exists</u> $r \supseteq q$ <u>with</u> $r \Vdash \varphi$.

To construct models by forcing we introduce the notion of a generic set. The following theory is due to Robinson [28].

DEFINITION. A <u>generic set</u> for T is a set G of basic sentences such that each finite $p \subseteq G$ is a condition for T and for each sentence φ of L , there is a $p \subseteq G$ which either forces φ or forces $\neg\varphi$. $G \Vdash \varphi$ means that some $p \subseteq G$ forces φ .

LEMMA 5.2. <u>For every condition</u> p <u>there is a generic set</u> G <u>containing</u> p .

GENERIC MODEL THEOREM. <u>For each generic set</u> G <u>for</u> T <u>there is a unique (up to isomorphism) model</u> $\underline{A}(G)$ <u>for</u> L <u>such that</u>

(i) <u>Each</u> $a \in A(G)$ <u>is the interpretation of some</u> $c \in C$.

(ii) <u>For each sentence</u> φ <u>in</u> L , $\underline{A}(G) \models \varphi$ <u>iff</u> $G \Vdash \varphi$.

DEFINITION. $\underline{A}$ is a T-<u>generic</u> <u>model</u> if $\underline{A} = \underline{A}(G)$ for some generic set G for T .

The proof of the Generic Model Theorem is by an induction on the complexity of formulas similar to the proof of the compactness theorem.

THEOREM 5.2. (i) $p \Vdash^{w} \varphi$ <u>if and only if</u> φ <u>holds in all</u> T-<u>generic models</u> $\underline{A}(G)$ <u>such that</u> $p \subseteq G$.

(ii) <u>Every T-generic model is a model of</u> T .

<u>Proof</u>. (i) is an easy consequence of the Generic Model Theorem. (ii) depends on our assumption that the axioms of T are $\forall\exists$. Consider an axiom of the form $\forall\vec{x}\exists\vec{y}\varphi(\vec{x},\vec{y})$, that is, $\neg\exists\vec{x}\neg\exists\vec{y}\varphi(\vec{x},\vec{y})$. If this axiom fails in some $\underline{A}(G)$, then some p forces $\exists\vec{x}\neg\exists\vec{y}\varphi(\vec{x},\vec{y})$, hence forces $\neg\exists\vec{y}\varphi(\vec{c},\vec{y})$ for some $\vec{c}$. But p is consistent with T , so some $q \supseteq p$ must contain $\varphi(\vec{c},\vec{d})$ for some $\vec{d}$. Then q forces $\exists\vec{y}\varphi(\vec{c},\vec{y})$, a contradiction. ⊣

The following theorems give a feeling for the nature of the T-generic models.

DEFINITION. A model $\underline{A}$ of T is <u>existentially closed</u> in T if every existential sentence which holds in some

extension of $\underline{A}_A$ which is a model of T holds in $\underline{A}_A$.
If the class of all existentially closed models in T is characterized by some theory T* in L , T* is called the model companion of T.

Examples: The theory of algebraically closed fields is the model companion of the theory of fields. The theory of real closed (ordered) fields is the model companion of the theory of formally real (ordered) fields. The theory of groups does not have a model companion. For a survey of this extensive subject see Hirschfield and Wheeler 10 .

THEOREM 5.3. *Every generic model for T is existentially closed in T.*

THEOREM 5.4. *If T has a model companion T*, then the class of all T-generic models is equal to the class of all countable models of T* .*

The proof of 5.3 is straightforward. The proof of 5.4 uses 5.3 and the fact that T* is model-complete, that is, if $\underline{A}$ is a model of T* and $\underline{A}_A \models \varphi$ then $T^* \cup D(\underline{A}) \models \varphi$.
To prove 5.4, let $\underline{A}$ be a countable model of T*, let G be the diagram of $\underline{A}$, and show by induction that if $\underline{A}_A \models \varphi$ then $G \Vdash \varphi$. It then follows that G is generic; $\underline{A}_A \cong \underline{A}(G)$.

LECTURE 6. MODEL THEORY FOR $\mathcal{L}_{\omega_1\omega}$.

When we go beyond classical first order logic, Lindstrom's theorem shows that we must do without either countable compactness or the downward Löwenheim-Skolem theorem. The logic $\mathcal{L}_{\omega_1\omega}$ is not countably compact but still has a well behaved model theory. This is because the forcing construction is available as a kind of substitute for the compactness theorem. Another logic with a well behaved model theory, which we shall not have time to develop here, is the logic $\mathcal{L}(Q_{\omega_1})$. This logic is interesting because it is countably compact but does not have Löwenheim number ω . Kim Bruce 6 has recently shown that the model theory of $\mathcal{L}(Q_{\omega_1})$ can also be based on a forcing construction.

Historically, there are two separate lines of development of the forcing construction in $\mathcal{L}_{\omega_1\omega}$, the model existence theorem of Smullyan and Makkai, and Robinson forcing. The relation between the two approaches and the generalization of forcing to $\mathcal{L}_{\omega_1\omega}$ is explained in Keisler 14 .

$\mathcal{L}_{\omega_1\omega}$ is like $\mathcal{L}_{\omega\omega}$ but allows the conjunction and disjunction of countable sets of formulas. As in the preceding lecture we regard $\neg$, $\vee$, and $\exists$ as fundamental. We assume in this lecture that L is countable and contains an infinite set C of constant symbols. Denote by $L_{\omega_1\omega}$ the set of all sentences φ of $\mathcal{L}_{\omega_1\omega}$ such that the symbols of φ are in L and only finitely many $c \in C$ occur in φ.

DEFINITION. By a <u>forcing base</u> we mean a countable subset S of $L_{\omega_1\omega}$ which contains all atomic sentences of L, and is closed under negations and under substitution instances of subformulas, where free variables are replaced by constants in C.

DEFINITION. Let S be a forcing base. By a <u>forcing property</u> on S we mean a nonempty set P of consistent finite subsets $p \subseteq S$ such that:

(i) If $p \in P$ and $q \subseteq p$ then $q \in P$.

(ii) If $p \models \varphi$ and $\varphi \in S$ then $p \cup \{\varphi\} \in P$.

(iii) If $\bigvee \Phi \in p$ then $p \cup \{\varphi\} \in P$ for some $\varphi \in \Phi$.

(iv) If $\exists u\, \varphi(u) \in p$ then $p \cup \{\varphi(c)\} \in P$ for some $c \in C$.

The elements $p \in P$ are called <u>conditions</u> of P. By (i), the empty condition $\emptyset$ always belongs to P.

DEFINITION. The relation p <u>forces</u> φ, in symbols $p \Vdash \varphi$ or $p \Vdash_P \varphi$, is between conditions p and arbitrary sentences $\varphi \in L_{\omega_1\omega}$. It is defined as follows.

If φ is atomic, then $p \Vdash \varphi$ iff $\varphi \in p$.

$p \Vdash \neg\varphi$ iff there is no $q \supseteq p$ in P with $q \Vdash \varphi$.

$p \Vdash \exists u\, \varphi(u)$ iff $p \Vdash \varphi(c)$ for some $c \in C$.

$p \Vdash \bigvee \Phi$ iff $p \Vdash \varphi$ for some $\varphi \in \Phi$.

P <u>weakly forces</u> φ, $p \Vdash^w \varphi$, if $p \Vdash \neg\neg\varphi$.

When we take S to be the set of all basic sentences of L and P the set of all finite subsets of S consistent with T, we obtain Robinson forcing. More generally, the set of all finite subsets of a forcing base S satisfiable in a class of models K is a forcing property. We shall use still other forcing properties later in this lecture.

The basic results on forcing are now familiar.

LEMMA 6.1. (i) <u>If</u> $p \subseteq q$ <u>and</u> $p \Vdash \varphi$ <u>then</u> $q \Vdash \varphi$.

(ii) <u>If</u> $\varphi \in p$ <u>then</u> $p \Vdash^w \varphi$.

(iii) $p \Vdash^w \varphi$ <u>iff for all</u> $q \supseteq p$ <u>there exists</u>

$r \supseteq q$ such that $r \Vdash \varphi$.

DEFINITION. A subset $G \subseteq S$ is said to be *generic* for P if:

(i) Each finite $p \subseteq G$ is a condition in P.

(ii) For each $\varphi \in S$ there exists $p \subseteq G$ such that $p \Vdash \varphi$ or $p \Vdash \neg\varphi$.

We write $G \Vdash \varphi$ if $p \Vdash \varphi$ for some $p \subseteq G$.

LEMMA 6.2. *Each condition* $p \in P$ *belongs to some generic set.*

GENERIC MODEL THEOREM. *For every generic set* G *for* P *there is a unique (up to isomorphism) model* $\underline{A}(G)$ *such that:*

(i) *Each element of* $A(G)$ *is the interpretation of some* $c \in C$.

(ii) *For each* $\varphi \in S$, $\underline{A}(G) \models \varphi$ *iff* $G \Vdash \varphi$.

We call $\underline{A}(G)$ the *generic model* generated by G.

We now give a variety of applications of forcing in $\mathcal{L}_{\omega_1\omega}$. These applications use the following countability lemma.

LEMMA 6.3. *Each countable set of sentences of* $L_{\omega_1\omega}$ *can be extended to a (countable) forcing base.*

There are 2^{ω} different sentences in $L_{\omega_1\omega}$. The lemma depends on the fact that L has countably many symbols and each single sentence has countably many subformulas.

DOWNWARD LÖWENHEIM-SKOLEM THEOREM. The logic $\mathcal{L}_{\omega_1\omega}$ has Löwenheim number ω.

Proof. Let φ be a consistent sentence in $L_{\omega_1\omega}$. Let S be a forcing base containing φ and let P be the set of all consistent finite subsets of S. By Lemma 6.2 and the Generic Model Theorem, φ holds in some generic model $\underline{A}(G)$. Since every element is an interpretation of some c in the countable set C, $\underline{A}(G)$ is countable.

OMITTING TYPES THEOREM. Let P be a forcing property on a forcing base S and for each $m,n<\omega$ and $c \in C$ let $\varphi_{mn}(c) \in S$. Suppose that for each $m<\omega$, $c \in C$, and $p \in P$, there exists n such that $p \cup \{\varphi_{mn}(c)\} \in P$. Then

$$\bigwedge_m \forall u \bigvee_n \varphi_{mn}(u)$$

has a model.

Proof. Using the definition of forcing, the above sentence is weakly forced by the empty condition, and thus holds in every generic model for P. ⊣

The following result is proved using ω_1 applications of the Omitting Types Theorem. An analogous theorem was earlier proved for $\mathcal{L}_{\omega\omega}$ by Vaught 33 using compactness.

TWO CARDINAL THEOREM. (Keisler 13). Suppose L has unary relations U, V, and a sentence $\varphi \in L_{\omega_1\omega}$ has a model $\underline{A}$ such that $\omega \leq |U^{\underline{A}}| < |V^{\underline{A}}|$. Then φ has a model $\underline{B}$ such that $|U^{\underline{B}}| = \omega$ and $|V^{\underline{B}}| = \omega_1$.

We shall omit the proof. The Two Cardinal Theorem can be used to prove the following extension of Theorem 1.2 to $\mathcal{L}_{\omega_1\omega}$. We call a sentence $\varphi \in L_{\omega_1\omega}$ κ-stable if for every model $\underline{A}$ of φ and every set $X \subseteq A$ of power at most κ, $\underline{A}$ has elements of at most κ different types over X (in the original sense of $\mathcal{L}_{\omega\omega}$).

THEOREM 6.4. If a sentence φ in $\mathcal{L}_{\omega_1\omega}$ is ω-stable, then φ is κ-stable for all κ. (Keisler 13).

Proof. Suppose φ is not κ-stable. Form a sentence ψ in an expansion of L such that if $(\underline{C}, U, V, \ldots)$ is a model of ψ, then $\underline{C} \models \varphi$ and the elements of V all have different types over U, and ψ has a model $(\underline{A}, U, V, \ldots)$ with $\omega \leq |U| < |V|$. Then ψ has a model $(\underline{B}, U', V', \ldots)$ with $|U'| = \omega$, $|V'| = \omega_1$. Hence $\underline{B}$ has ω_1 types over U', so φ is not ω-stable. ⊣

NON-WELLORDERING THEOREM. (Lopez-Escobar 19) Let $<$ be a binary relation symbol in L and suppose $\varphi \in L_{\omega_1 \omega}$ has a model $\underline{A}$ in which $<$ is a well ordering of type at least ω_1. Then φ has a model $\underline{B}$ in which $<$ is not a well ordering.

Proof. We may assume $\langle \omega_1, < \rangle \subseteq \underline{A}$. Let d_r, r rational, be new constant symbols added to L, and let S be a forcing base containing φ. Define P as follows. Let $p = p(\vec{c}, d_{r_1}, \ldots, d_{r_n})$ be a finite subset of S with $r_1 < r_2 < \ldots < r_n$. Then $p \in P$ if and only if for all $\alpha < \omega_1$, there are $\vec{a} \in A^m$ and ordinals $\beta_1, \ldots, \beta_n < \omega_1$ such that

$$\alpha \leq \beta_1, \quad \beta_1 + \alpha \leq \beta_2, \quad \ldots, \quad \beta_{n-1} + \alpha \leq \beta_n,$$

$$(\underline{A}, \vec{a}, \beta_1, \ldots, \beta_n) \models p .$$

P is a forcing property and $\varphi \in P$, so φ has a generic model $\underline{B} = \underline{A}(G)$. In $\underline{B}$, the interpretation of the d_r's are ordered like the rationals, so $<$ is not a well ordering in $\underline{B}$. ⊣

INTERPOLATION THEOREM. (Lopez-Escobar 18). Let $L' \cap L'' = L$. If φ, ψ are sentences in the languages $L'_{\omega_1 \omega}$ and $L''_{\omega_1 \omega}$ respectively and if $\models \varphi \rightarrow \psi$, then there

<u>is a sentence</u> θ <u>of</u> $L_{\omega_1\omega}$ <u>such that</u> $\models \varphi \to \theta$ <u>and</u> $\models \theta \to \psi$.

<u>Proof</u>. Assume no such θ exists. Let S' and S'' be forcing bases containing φ and ψ in $L'_{\omega_1\omega}$ and $L''_{\omega_1\omega}$. Let $p \in P$ if and only if $p = p' \cup p''$ where $p' \subseteq S'$, $p'' \subseteq S''$, p' is consistent with φ, p'' is consistent with $\neg\psi$, and there is no $\theta \in L_{\omega_1\omega}$ with $p' \models \theta$, $p'' \models \neg\theta$. Then P is a forcing property in a slightly more general sense, and $\varphi \wedge \neg\psi$ holds in a generic model $\underline{A}(G)$ for P. This contradicts the hypothesis

<u>Note</u>. In $\mathcal{L}_{\omega\omega}$ the Interpolation Theorem was first proved independently by Craig and Robinson, and is equivalent to the Robinson Consistency Theorem. However, in $\mathcal{L}_{\omega_1\omega}$ the analogue of the Robinson Consistency Theorem is false even though the Interpolation theorem is true.

H. J. Keisler

LECTURE 7. MODEL THEORY FOR $\mathcal{L}_{\infty\omega}$.

In this lecture we discuss some model theoretic constructions which are important in classical model theory and are available even in $\mathcal{L}_{\infty\omega}$. The language $\mathcal{L}_{\infty\omega}$ allows conjunctions and disjunctions over arbitrary sets of formulas. The forcing construction is no longer available because it depends on a sentence having countably many subformulas. The constructions still available are elementary chains, Skolem functions, and indiscernibles.

We fix a set L of symbols and let $L_{\infty\omega}$ be the set of all formulas φ of $\mathcal{L}_{\infty\omega}$ such that φ has finitely many free variables and the symbols of φ belong to some finite $L_0 \subseteq L$. The notion of a <u>subformula</u> of φ is defined in the natural way. Since $L_{\infty\omega}$ is a proper class, we need a way of choosing nice subsets of $L_{\infty\omega}$.

DEFINITION. A <u>fragment</u> of $L_{\infty\omega}$ is a subset S of $L_{\infty\omega}$ closed under subformulas, finitary connectives and quantifiers, and substitution of terms for free variables.

Clearly, each subset of $L_{\infty\omega}$ generates a fragment of $L_{\infty\omega}$.

DEFINITION. Let S be a fragment of $L_{\infty\omega}$. $\underline{A}$ is an S-<u>elementary submodel</u> of $\underline{B}$, in symbols $\underline{A} \prec_S \underline{B}$, if $\underline{A} \subseteq \underline{B}$ and for every $\varphi \in S$ and $\vec{a}$ in A we have $(\underline{A},\vec{a}) \models \varphi$ if and only if $(\underline{B},\vec{a}) \models \varphi$.

$\underline{A}_\alpha$, $\alpha < \gamma$, is an S-<u>elementary chain</u> if $\underline{A}_\alpha \prec_S \underline{A}_\beta$ whenever $\alpha < \beta < \gamma$.

ELEMENTARY CHAIN THEOREM. (Tarski and Vaught 32). <u>If $\underline{A}$ is the union of an S-elementary chain $\underline{A}_\alpha$, $\alpha < \gamma$, then each $\underline{A}_\alpha$ is an S-elementary submodel of $\underline{A}$.</u>

The proof is by induction on the complexity of formulas, and depends on the fact that each subformula has finitely many free variables.

The elementary chain theorem is often used together with other methods. For example, it is needed in the proof of the Two Cardinal Theorem in the last lecture.

We now introduce Skolem functions in $L_{\infty\omega}$.

DEFINITION. Let S be a fragment of $L_{\infty\omega}$. The <u>Skolem expansion</u> of S is constructed as follows. For each formula $\varphi(u,\vec{v}) \in S$ add a new function symbol $F_\varphi(\vec{v})$ to L, forming L^1 . Let S^1 be the fragment of L^1 generated by S . Then iterate the procedure, forming

$$L, L^1, L^2, \ldots, \quad S, S^1, S^2, \ldots \quad .$$

The unions $L^* = \bigcup_n L^n$ and $S^* = \bigcup_n S^n$ are called the S-Skolem expansions of L and S . The S-Skolem theory T_{Skolem} in $L^*_{\infty\omega}$ is the theory with the axioms

$$\exists u \varphi(u,\vec{v}) \rightarrow \varphi(F_\varphi(\vec{v}), \vec{v}) \quad , \quad \varphi \in S^* .$$

Notice that each axiom of T_{Skolem} belongs to S^* . Models of T_{Skolem} are called S-Skolem models .

We state three classical applications of Skolem functions.

SKOLEM EXPANSION THEOREM. (i) Every model $\underline{A}$ for L has an expansion to an S-Skolem model $\underline{A}^*$.

(ii) If $\underline{A}^*$ is an S-Skolem model, i.e. a model of T_{Skolem} , then every submodel $\underline{B}^* \subseteq \underline{A}^*$ is an S^*-elementary submodel of $\underline{A}^*$, $\underline{B}^* \prec_{S^*} \underline{A}^*$.

DOWNWARD LOWENHEIM-SKOLEM THEOREM. (Hanf 9) If S has power κ , then every model $\underline{A}$ has an S-elementary submodel of power at most κ .

DEFINITION. Given a fragment S and model $\underline{A}$, let $Th_S(\underline{A})$, the S-theory of $\underline{A}$, be the set of all sentences $\varphi \in S$ which hold in $\underline{A}$.

PATCHING THEOREM. Let T^* be a theory in the Skolem fragment S^* of L^* such that for each finite subset

$L_0 \subseteq L$ there is an S_0-Skolem model $\underline{B}^*$ whose S_0^* theory is $T^* \cap S_0^*$. Then T^* has an S-Skolem model $\underline{A}^*$.

Proof. Build $\underline{A}^*$ out of terms in L^* in such a way that an atomic sentence holds in $\underline{A}^*$ if and only if it is in T^*. ⊣

We now turn to the notion of indiscernibles. They were introduced by Ehrenfeucht and Mostowski[8] for $\mathcal{L}_{\omega\omega}$, and have been an important tool in both finitary and infinitary logic.

DEFINITION. Let X be a subset of A and < a linear order on X. Given a fragment S and a model $\underline{A}$, we say that $\langle X, < \rangle$ is S-indiscernible in $\underline{A}$ if any two increasing n-tuples $x_1 < \ldots < x_n$ and $y_1 < \ldots < y_n$ from $\langle X, < \rangle$ satisfy the same formulas of S in $\underline{A}$.

EHRENFEUCHT-MOSTOWSKI THEOREM. Every theory T in $L_{\omega\omega}$ which has an infinite model has a Skolem model with an infinite set of $L^*_{\omega\omega}$-indiscernibles.

The Ehrenfeucht-Mostowski Theorem is a special case of the following theorem for infinitary logic.

STRETCHING THEOREM. Let φ belong to a fragment S of $L_{\infty\omega}$. The following are equivalent.

(i) φ has models of arbitrarily large cardinality.

(ii) φ *has an S-Skolem model* $\underline{A}^*$ *with an infinite set of* S^*-*indiscernibles*.

Proof. Assume (ii) and let $\langle X,<\rangle$ be an infinite set of S^*-indiscernibles in $\underline{A}^*$. Given any cardinal κ let $\langle Y,<\rangle$ be a linearly ordered set of power κ. Form the expansion L' of L* by adding a new constant symbol c_y for each $y \in Y$, and let S' be the fragment of L' generated by S*. Let T' be the theory in S' such that $\psi(c_{y_1},\ldots,c_{y_n}) \in T'$ if and only if $(\underline{A}^*,x_1,\ldots,x_n)$ satisfies ψ for all $x_1 < \ldots < x_n$ in $\langle X,<\rangle$. By the Patching Theorem, T' has a model B^*_Y, and the reduct $\underline{B}$ is a model of φ of power at least κ. Thus (ii) implies (i). The converse is harder and also uses the Patching Theorem, along with the Erdös-Rado theorem on infinite partitions. ⊣

NON-WELLORDERING THEOREM FOR $\mathcal{L}_{\infty\omega}$ (Lopez-Escobar 19). *Let* $<$ *be a binary relation symbol in* L. *Suppose* $\varphi \in L_{\infty\omega}$ *and for each ordinal* α, φ *has a model in which* $<$ *is a well ordering of type at least* α. *Then* φ *has a model in which* $<$ *is not a well ordering*.

Proof. Let S be a fragment containing φ. φ has arbitrarily large models, so by the Stretching Theorem φ has a Skolem model $\underline{A}^*$ with an infinite set of S^*-indiscernibles. Using the proof of the Stretching Theorem with Y ordered like the rationals, we obtain a model $\underline{B}^*$ in which $<$ is not a well ordering. $\dashv$

Morley used indiscernibles to compute the Hanf number of $\mathcal{L}_{\omega_1\omega}$ and several related logics. The cardinals $\beth_\alpha$ are defined recursively as follows:

$$\beth_0 = \omega \;,\quad \beth_{\alpha+1} = 2^{\beth_\alpha} \;,\quad \beth_\alpha = \bigcup_{\beta<\alpha} \beth_\beta \quad \text{for limit } \alpha .$$

THEOREM 7.1. (Morley 23). <u>The Hanf number of $\mathcal{L}_{\omega_1\omega}$ is $\beth_{\omega_1}$</u> .

<u>Proof</u>. To show that the Hanf number is at least $\beth_\alpha$, we need an example of a sentence φ_α of $\mathcal{L}_{\omega_1\omega}$ which has a model of power at least $\beth_\alpha$ but does not have arbitrarily large models. Let $\alpha < \omega_1$. The first step is to construct a sentence which has a well ordered model $\underline{B}_\alpha$ of order type α . This is easily done by induction on α . The final sentence φ_α describes an $\in$ structure in which every set has rank less that α .

To show that the Hanf number is at most $\beth_{\omega_1}$, suppose $\varphi \in L_{\omega_1\omega}$ has a model of power at least $\beth_{\omega_1}$. Let S be a countable fragment of $L_{\omega_1\omega}$ containing φ. By a proof similar to the Non-wellordering Theorem for $\mathcal{L}_{\omega_1\omega}$ it can be shown that φ has an S-Skolem model $\underline{A}^*$ with an infinite set of S^*-indiscernibles. By the Stretching Theorem, φ has arbitrarily large models. ⊣

LECTURE 8. INFINITARY SOFT MODEL THEORY.

In this lecture we return to soft model theory in the light of the methods developed in the preceding two lectures. We shall give an analogue of Lindstrom's theorem for $\mathcal{L}_{\infty\omega}$ due to Barwise 3, which shows that $\mathcal{L}_{\infty\omega}$ is a "natural" logic. We introduce an abstract Skolem expansion property which has applications even beyond $\mathcal{L}_{\infty\omega}$. Then we discuss the open problem of characterizing $\mathcal{L}_{\omega_1\omega}$ as a largest logic. In our discussion we shall examine the relationship between several familiar properties of $\mathcal{L}_{\omega_1\omega}$ in the abstract setting. It should be mentioned that Makowsky, Shelah, and Stavi 21 and others have begun a more ambitious project for soft model theory, to find new "natural" logics.

For brevity we denote an abstract logic $(\mathcal{L}^*, \models^*)$ by $\mathcal{L}^*$. If $\underline{A}$ is a model for L, let $Th_{\mathcal{L}^*}(\underline{A}) = \{\varphi \in \mathcal{L}^* : L_\varphi \subseteq L \text{ and } \underline{A} \models^* \varphi\}$. We call $\mathcal{L}^\circ$ a <u>sublogic</u> of $\mathcal{L}^*$ if $\mathcal{L}^\circ \subseteq \mathcal{L}^*$ and $\models^\circ$ agrees with $\models^*$ on $\mathcal{L}^\circ$. $\mathcal{L}^\circ$ is <u>weaker than</u> $\mathcal{L}^*$, $\mathcal{L}^\circ \leq \mathcal{L}^*$, if there is a mapping $f: \mathcal{L}^\circ \to \mathcal{L}^*$ such that each $\varphi \in \mathcal{L}^\circ$ has exactly the same models as $f\varphi$.

REMARK. $\mathcal{L}^\circ \subseteq \mathcal{L}^*$ implies that $\mathcal{L}^\circ \leq \mathcal{L}^*$. $\mathcal{L}^\circ \leq \mathcal{L}^*$ implies that for some $\mathcal{L}' \subseteq \mathcal{L}^*$, $\mathcal{L}' \leq \mathcal{L}^\circ \leq \mathcal{L}'$.

Here is a list of properties which hold at least for every logic $\mathcal{L}^* \subseteq \mathcal{L}_{\infty\omega}$.

KARP PROPERTY. If $\underline{A} \cong_p \underline{B}$ then $\underline{A} \equiv_{\mathcal{L}^*} \underline{B}$ (Lecture 4).

By a <u>well ordered model</u> we mean a model $\underline{A}$ in which $<^{\underline{A}}$ is a well ordering.

NON WELLORDERING PROPERTY. Any $\varphi \in \mathcal{L}^*$ with well ordered models of arbitrarily large order type has a non well ordered model.

$\mathcal{L}^*$ is said to be <u>locally bounded</u> if for each L, the class $\{\varphi : L_\varphi \subseteq L\}$ is a set. Note that $\mathcal{L}_{\infty\omega}$ is not locally bounded, but each $\varphi \in \mathcal{L}_{\infty\omega}$ belongs to a locally bounded sublogic.

SKOLEM EXPANSION PROPERTY. (SEP). Each $\varphi \in \mathcal{L}^*$ belongs to a locally bounded sublogic $\mathcal{L}^{\circ} \subseteq \mathcal{L}^{\lambda}$ with the following properties. For each L there is associated a language L^{+} and a class of models $K(L)$ for L^{+} called <u>Skolem models</u> for L , such that:

(i) $L^{+} = \bigcup\{L_0^{+} : L \supseteq L_0 \text{ finite}\}$, and $\underline{A} \in K(L)$ iff $\underline{A}\restriction L_0^{+} \in K(L_0)$ for all finite $L_0 \subseteq L$.

(ii) $(L_X)^{+} = (L^{+})_X$ where X is a set of constants.

(iii) Each model $\underline{A}$ for L can be expanded to a Skolem model for L , i.e. a model $\underline{B} \in K(L)$.

(iv) (Patching). Let $T \subseteq \mathcal{L}^{\circ}$. If for every finite $L_0 \subseteq L$ there is a Skolem model $\underline{A}_0$ for L_0 with $Th_{\mathcal{L}^{\circ}}(\underline{A}_0) \subseteq T$, then there is a Skolem model $\underline{A}$ for L with $Th_{\mathcal{L}^{\circ}}(\underline{A}) \subseteq T$.

THEOREM 8.1. (Karp 11). $\underline{A} \cong_p \underline{B}$ iff $\underline{A} \equiv_{\mathcal{L}_{\infty\omega}} \underline{B}$.

Proof. If $\underline{A} \equiv_{\mathcal{L}_{\infty\omega}} \underline{B}$ then $I: \underline{A} \cong_p \underline{B}$ where $\vec{a} I \vec{b}$ iff $(\underline{A}, \vec{a}) \equiv_{\mathcal{L}_{\infty\omega}} (\underline{B}, \vec{b})$. If $\underline{A} \cong_p \underline{B}$ then it can be shown by induction on φ that $\vec{a} I \vec{b}$ implies $(\underline{A}, \vec{a}) \models \varphi$ iff $(\underline{B}, \vec{b}) \models \varphi$.

THEOREM 8.2. (Barwise 3). $\mathcal{L}_{\infty\omega}$ <u>is the strongest logic with the Karp and Non Wellordering properties</u>.

The proof is like the proof of Lindstrom's Theorem and uses Theorem 8.1. This theorem characterizes $\mathcal{L}_{\infty\omega}$.

By the results of Lecture 7 , $\mathcal{L}_{\infty\omega}$ has the SEP.

The SEP is fundamental and quite powerful. Fuhrken 34 essentially showed that $\mathcal{L}(Q_\kappa)$ has the SEP. In fact, it can be shown that every sublogic of the enormous logic $\mathcal{L}(Q_\kappa : \kappa$ a cardinal$)$ has the SEP.

THEOREM 8.3. (Barwise and Keisler). <u>If</u> $\mathcal{L}^*$ <u>has the</u> SEP,

(i) <u>The natural analogue of the Stretching Theorem holds in</u> $\mathcal{L}^*$.

(ii) $\mathcal{L}^*$ <u>has the Non Wellordering Property</u>.

The proof is like the case for $\mathcal{L}_{\infty\omega}$ in Lecture 7.

COROLLARY 8.4. $\mathcal{L}_{\infty\omega}$ <u>is the strongest logic with the Karp Property and the SEP</u>.

When $\kappa = \beth_\kappa$, the logics $\mathcal{L}_{\kappa\omega}$ can be characterized in a way analogous to the above characterizations of $\mathcal{L}_{\infty\omega}$ (Barwise 3). $\mathcal{L}_{\kappa\omega}$ is the sublogic of $\mathcal{L}_{\infty\omega}$ consisting of all sentences φ with fewer than κ subformulas.

The more interesting problem of characterizing $\mathcal{L}_{\omega_1\omega}$ as a strongest logic remains open. This is surprising since so much is known about $\mathcal{L}_{\omega_1\omega}$. Here are some abstract properties which hold for $\mathcal{L}_{\omega_1\omega}$.

ω_1-NON WELLORDERING PROPERTY. If $\varphi \in \mathcal{L}^*$ has a well ordered model of order type at least ω_1, then φ has a non well ordered model.

INTERPOLATION PROPERTY. If $\models^* \varphi \to \psi$, there is a sentence θ such that $L_\theta \subseteq L_\varphi \cap L_\psi$ and $\models^* \varphi \to \theta$, $\models^* \theta \to \psi$. ($\models^* \varphi$ means that $\underline{A} \models^* \varphi$ for all models $\underline{A}$ for L_φ).

OMITTING TYPES PROPERTY. Every $\varphi \in \mathcal{L}^*$ belongs to a locally bounded $\mathcal{L}^\circ \subseteq \mathcal{L}^*$ such that for each countable L and extra constant symbol $c \notin L$, the following holds.

Let $T \subseteq \mathcal{L}^\circ$ be a consistent set of sentences of L and $\varphi_{mn} \in \mathcal{L}^\circ$, $m, n < \omega$, be sentences of $L \cup \{c\}$. Suppose that for each sentence $\psi \in \mathcal{L}^\circ$ of $L \cup \{c\}$ which is consistent with T, and each m, there exists n such that $\psi \wedge \varphi_{mn}$ is consistent with T. Then T has a model $\underline{A}$ such that

$$(\forall m < \omega)(\forall a \in A)(\exists n < \omega)(\underline{A}, a) \models^* \varphi_{mn} .$$

To this list we may add the Karp and SEP Properties, Löwenheim number ω, and Hanf number $\beth_{\omega_1}$.

Kunen has recently given an example (assuming the continuum hypothesis) of a logic $\mathcal{L}^*$ which is properly stronger than $\mathcal{L}_{\omega_1\omega}$ and has all the above properties. This suggests that $\mathcal{L}_{\omega_1\omega}$ is not a natural logic after all, and the real problem is to find the strongest logic which behaves like $\mathcal{L}_{\omega_1\omega}$. Kunen's example adds a new infinite connective to $\mathcal{L}_{\omega_1\omega}$. We conclude by stating some results which give relationships between the above properties.

THEOREM 8.5. (Barwise 3). <u>If</u> $\mathcal{L}^*$ <u>has the Karp Property and the Interpolation Property, then</u> $\mathcal{L}^*$ <u>has Löwenheim number</u> ω .

THEOREM 8.6. (Barwise 3). <u>If</u> $\mathcal{L}^*$ <u>has Löwenheim number</u> $\leq \beth_{\omega_1}$ <u>and Hanf number</u> $\leq \beth_{\omega_1}$ <u>then</u> $\mathcal{L}^*$ <u>has the</u> ω_1<u>-non Wellordering Property.</u> (In fact, this result holds for any infinite ordinal α in place of ω_1).

THEOREM 8.7. (Krynicki and Onyszkiewicz). <u>If</u> $\mathcal{L}^*$ <u>has the Omitting Types Property, then</u> $\mathcal{L}^*$ <u>has Löwenheim number</u> ω .

The following result is new.

THEOREM 8.8. <u>Assume the Axiom of Constructibility, or at least that there is a subset</u> r $\subseteq \omega$ <u>such that uncountably many subsets of</u> ω <u>are constructible from</u> r ; <u>If</u> $\mathcal{L}^*$ <u>has the Omitting Types Property, then</u> $\mathcal{L}^*$ <u>has the</u> ω_1 <u>-non Wellordering Property.</u>

REFERENCES

1. K.J.Barwise. Absolute logics and $L_{\infty\omega}$. AML 4 (1972) pp. 309-340.

2. K.J.Barwise. Back and forth through infinitary logic. pp. 5-34 in Morley 24.

3. K.J.Barwise. Axioms for abstract model theory. AML 7 (1974), 221-265.

4. K.J.Barwise and H.K.Kunen. Hanf numbers for fragments of $L_{\infty\omega}$. Israel J. Math. 10 (1971), pp. 306-320.

5. K.J.Barwise and J.Schlipf. Recursively saturated and resplendent models. To appear.

6. K. Bruce. Model-theoretic forcing and $L(\exists)$. Thesis, U. of Wisconsin, 1975.

7. C.C.Chang and H.J.Keisler. Model Theory. North-Holland 1973.

8. A. Ehrenfeucht and A. Mostowski. Models of axiomatic theories admitting automorphisms. Fund Math. 43 (1956) pp. 50-68.

9. W. Hanf. Incompactness in languages with infinitely long expressions. Fund. Math. 53 (1964), pp. 309-324.

10. J. Hirschfeld and W.H.Wheeler. Forcing, arithmetic and division rings. To appear.

11. C. Karp. Finite quantifier equivalence. Pp. 407-412 in The Theory of Models, ed. by Addison, Henkin, and Tarski, North-Holland 1965.

12. C. Karp, Languages with expressions of infinite length. North-Holland 1964.

13. H.J.Keisler. Model theory for infinitary logic. North-Holland 1971.

14. H.J.Keisler. Forcing and the omitting types theorem. Pp. 96-133 in Morley 24.

15. H.J.Keisler. Logic with the quantifier "there exist uncountably many". AML 1 (1970), pp. 1-93.

16. M.Makkai. Preservation theorems for logic with denumerable conjunctions and disjunctions. JSL 34 (1969), pp. 437-459.

17. P.Lindstrom. On extensions of elementary logic. Theoria 35 (1969), pp. 1-11.

18. E. Lopez-Escobar. An interpolation theorem for denumerably long sentences. Fund. Math. 58 (1965), 253-277.

19. E. Lopez-Escobar. On definable well orderings. Fund. Math. 58 (1966), pp. 13-21.

20. R. Lyndon. Properties preserved under homomorphism. Pacific J. Math. 9 (1959), pp. 143-154.

21. J.A.Makowsky, S. Shelah, and J. Stavi. Δ-logics and generalized quantifiers. To appear.

22. M.Morley. Categoricity in power. TAMS 114 (1965), pp. 514-538.

23. M.Morley. Omitting classes of elements. Pp. 265-273 in The Theory of Models, ed. by Addison, Henkin and Tarski, North-Holland 1965.

24. M. Morley. Studies in model theory. Math. Assn. of Amer. 1973.

25. M. Presburger. Warsaw 1930.

26. A. Robinson. A result on consistency and its application to the theory of definition. Indag. Math. 18 (1956), pp. 47-58.

27. A. Robinson. Introduction to model theory and to the metamathematics of algebra. North-Holland 1963.

28. A. Robinson. Forcing in model theory. Symp. Math. 5 (1971), pp. 69-82.

29. S. Shelah. Stability, the finite cover property, and superstability. AML 3 (1971), pp. 271-362.

30. J. Schlipf. Some hyperelementary aspects of model theory. Thesis, U. of Wisconsin, 1975.

31. A. Tarski and J.C.C.McKinsey. A decision method for elementary algebra and geometry. Rand Corp. 1948.

32. A. Tarski and R. Vaught. Arithmetical extensions of relational systems. Comp. Math. 13 (1957), pp. 81-102.

33. M.Morley and R. Vaught. Homogeneous universal models. Math. Scand. 11 (1962), pp. 37-57.

34. G. Fuhrken. Skolem-type normal forms for a first-order language with a generalized quantifier. Fund. Math. 54 (1964), pp. 291-302.

CENTRO INTERNAZIONALE MATEMATICO ESTIVO

(C.I.M.E.)

SH - FORMULAS UND GENERALIZED EXPONENTIAL

M. SERVI

Corso tenuto a Bressanone dal 20 al 28 giugno 1975

MARIO SERVI

SH-Formulas and Generalized Exponential

(Lecture delivered on June 25, 1975)

There are two ideas underlying this talk of mine. The first one is concerned with the interpretation of a class of first order formulas (called SH-formulas, since they are a class of HORN formulas in SKOLEM open form) in a category satisfying a very weak requirement, namely the sole existence of finite products. This idea was developed some years ago and later I will devote some of my time to discuss it. Of course, it is fashionable to interpret *every* formula (even of higher order and multi-sorted) in a topos, but topoi have a very rich structure and satisfy many more properties than just having finite products. I might mention that C. MARCHINI started working on the comparison between the two definitions of truth and will soon publish a paper on the subject (see [1, 2]).

In order to explain the second idea, I have to recall the well known fact that the set-theoretic exponentiation (Cartesian power) splits into three non-equivalent concepts for general categories: 1) the Hom-bifunctor (any category $\underline{\underline{C}}$, Hom: $\underline{\underline{C}}^{o} \times \underline{\underline{C}} \longrightarrow \underline{\underline{S}}$), 2) the LAWVERE exponential (Cartesian closed category $\mathbf{C}$, Exp: $\mathbf{C}^{\circ}\times\mathbf{C} \longrightarrow \mathbf{C}$) and 3) the S-fold product of an object by itself (when such exists, S an object of the category $\mathbf{S}$ of sets), hereafter indicated by $\prod$: $\mathbf{S}^{\circ}\times\mathbf{C} \longrightarrow \mathbf{C}$. Now, the idea is to find a suitable generalization of these three concepts which will behave like them as far as the truth of SH's is concerned. In a previous work of mine ([5]) I introduced a bifunctor G: $\mathbf{D}^{\circ}\times\mathbf{D} \longrightarrow \mathbf{C}$ ($\mathbf{C}$, $\mathbf{D}$ suitable categories) satisfying hypotheses which hold for both Hom and Exp, but failed to include $\prod$ as a special case, since $\mathbf{D}$ is not, in general, the category of sets. Let us summarize all this as follows:

0) $(-)^{(-)}$: $\mathbf{S}^{\circ}\times\mathbf{S} \longrightarrow \mathbf{S}$; starting case

1) Hom: $\mathbf{C}^{\circ}\times\mathbf{C} \longrightarrow \mathbf{S}$

2) Exp: $\mathbf{C}^{\circ}\times\mathbf{C} \longrightarrow \mathbf{C}$ — the three generalizations of 0) (cases 1)–3));

3) $\prod$: $\mathbf{S}^{\circ}\times\mathbf{C} \longrightarrow \mathbf{C}$

4) G: $\mathbf{D}^{\circ}\times\mathbf{D} \longrightarrow \mathbf{C}$, introduced in [5];

5) (?): $\mathbf{E}^{\circ}\times\mathbf{D} \longrightarrow \mathbf{C}$, wanted.

The bifunctor (?): $\mathbf{E}^{\circ}\times\mathbf{D} \longrightarrow \mathbf{C}$ of case 5), although more general than the previous ones, should satisfy enough assumptions as to yield the usual result which holds for *direct powers*:

Let $\mathcal{B}$ be any $\mathcal{L}$-structure (in $\mathbf{D}$) and let H be any SH. Then H is true in $\mathcal{B}$ iff it is "uniformly true" in the $G(X, \mathcal{B})$'s (induced structures) with $X \in \mathrm{Ob}\ \underline{\underline{E}}$. In formulas:

$$\models_{\mathcal{B}} H \quad (\text{in } \underline{\underline{D}}) \qquad \text{iff} \qquad \models^{X \in \underline{\underline{E}}}_{G(X,\mathcal{B})} H \quad (\text{in } \underline{\underline{C}}).$$

Now, a few words of explanation about SH-formulas and their satisfaction. I will not spell out how SH's are made (see [4] for details), but I will give a general idea, pointing out the main features. First of all, I will confine myself to a first order language $\mathcal{L}$ with no function nor constant symbols. The expanded language $\mathcal{L}^*$, obtained by expanding $\mathcal{L}$ with dummy function symbols, will have the auxiliary purpose of taking care of existential quantifiers. As it is customary when dealing with interpretations of formulas in a category, we are interested in ordered pairs (H, n) - H a formula and $\underline{n}$ a natural number - rather than in formulas themselves. The natural number component will be taken care of by incorporating it inside the formula, so that an SH will not be directly recognizable as a first order formula of the ordinary kind. The main difference, though, will be just the following minor change: the supply of variables will be a triangular array

$$\varepsilon^n_1, \ \varepsilon^n_2, \ \ldots, \ \varepsilon^n_n \qquad (n = 1, 2, \ldots).$$

They will be inteppreted as projections, $\varepsilon^n_k : A^n \longrightarrow A$ being the kth n-place projection on A, the universe of interpretation.

The natural number associated with an SH will be called its <u>rank</u>. A typical atomic SH will have the form $P(t_1, \ldots, t_m)$ with terms $t_1, \ldots, t_m$ built up with dummy function symbols. An <u>elementary SH</u> will be of the form $H_1 \wedge \ldots \wedge H_r \longrightarrow H_0$, with $0 \leqslant r$, H_k atomic (fixed rank), and a general SH will be a conjunction of elementary SH's of the same rank.

Now for truth. First of all, an $\mathcal{L}$-<u>structure</u>, in a category $\underline{\underline{C}}$ with finite products, will be the usual thing, i. e. a pair $\mathcal{A} = (A, \Psi)$, with A an object of $\underline{\underline{C}}$ (the <u>carrier</u> or universe) and Ψ the "structure" (realizing the predicate symbols), bearing in mind that an m-ary relation on A is just a subobject $R \overset{u}{\rightarrowtail} A^m$ (no characteristic functions are available). Similarly for $\mathcal{L}^*$. An $\mathcal{L}^*$-structure $\mathcal{A}^* = (A, \Psi^*)$ will also be called an $\mathcal{L}$-<u>interpretation</u>, and it will be said to be <u>associated</u> with a structure $\mathcal{A}$, if $\mathcal{A}$ is the reduct to $\mathcal{L}$ of $\mathcal{A}^*$.

Note that interpretations serve the only purpose of defining truth, according to the following definition:

$\models_{\mathcal{A}} H$ (in $\underline{\underline{C}}$) if there is an $\mathcal{A}^*$ associated with $\mathcal{A}$ such that $\models_{\mathcal{A}^*} H$. Truth in $\mathcal{A}^*$, in turn, will be given in terms of satisfacion, which I will define explicitly only for atomic SH's. Thus, let H_0 be atomic of rank n, say $H_0 = P(t_1, \ldots, t_m)$, let $R \overset{u}{\rightarrowtail} A^m$ be $P^{\mathfrak{A}} = P^{\mathfrak{A}^*}$ and let $g\colon X \longrightarrow A^n$. We will write $\models_{\mathfrak{A}^*} H_0[g]$ (in $\underline{\underline{C}}$), if there is a morphism $X \longrightarrow R$ such that the following commutes:

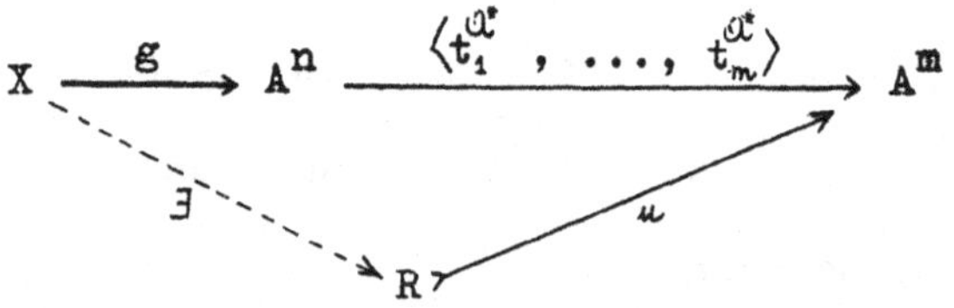

The definition extends to arbitrary SH's in the usual way. Finally, $\models_{\mathfrak{A}^*} H$ means $\models_{\mathfrak{A}^*} H[g]$, for every X and every $g\colon X \longrightarrow A^n$.

Let us now introduce the "generalized exponential". Let $\underline{\underline{E}}$, $\underline{\underline{D}}$, $\underline{\underline{C}}$ be three categories, $\underline{\underline{D}}$ with finite products. $\underline{\underline{C}}$, $\underline{\underline{D}}$ are not required to have a l l finite products, but they will turn out to have enough of them for the interpretations of symbols in $\mathcal{L}^*$ to make sense. Let $G\colon \underline{\underline{E}}^o \times \underline{\underline{D}} \longrightarrow \underline{\underline{C}}$ be a bifunctor, $U \in \mathrm{Ob}\ \underline{\underline{C}}$, $K\colon \underline{\underline{D}} \longrightarrow \underline{\underline{E}}$ a functor such that the following axioms hold:

(i) the standard functor $\underline{\underline{C}}[U, -]\colon \underline{\underline{C}} \longrightarrow \underline{\underline{S}}$ is full and faithful;

(ii) there is a natural bijection

$$\Phi_{-,-}\colon \underline{\underline{E}}[-, K(-)] \xrightarrow{\ \simeq\ } \underline{\underline{C}}[U, G(-, -)],$$

i. e. a bijection:

$$\frac{X \longrightarrow K(Y)}{U \longrightarrow G(X,\ Y)}$$

natural both in X and Y, $X \in \mathrm{Ob}\ \underline{\underline{E}}$, $Y \in \mathrm{Ob}\ \underline{\underline{D}}$, the upper arrow in $\underline{\underline{E}}$ and the lower one in $\underline{\underline{C}}$;

(iii) K is full and faithful and has a left adjoint, $F \dashv K$.

It is an easy matter to check that, with a suitable choice of K and U and enough hypotheses on $\underline{\underline{C}}$, Hom, Exp and $\prod$ meet these requirements.

It turns out to be useful to consider the functor

$$\hat{G}\colon \underline{\underline{D}} \longrightarrow \underline{\underline{C}}^{\underline{\underline{E}}^o}$$

given by $\hat{G}(x)(y) = G(y, x)$. In case G is Hom, then $\hat{G}$ is the YONEDA enbedding; the following theorem states that $\hat{G}$ turns out to be an enbedding in this general case as well.

THEOREM 1. - <u>$\hat{G}$ is full and faithful</u>.

THEOREM 2. - <u>Let $\underline{\underline{I}}$ be a diagram schema. If K preserves $\underline{\underline{I}}$-limits, then for every $X \in \mathrm{Ob}\ \underline{\underline{E}}$, $G(X, -)$ preserves $\underline{\underline{I}}$-limits</u>.

COROLLARY 1. - <u>For every $X \in \mathrm{Ob}\ \underline{\underline{E}}$, $G(X, -)$ preserves limits</u>.

COROLLARY 2. - <u>$\hat{G}$ preserves finite products and monos</u>.

Now, corollary 2 permits us to associate, with every structure $\mathcal{B} = (B, \Psi)$ (in $\underline{\underline{D}}$) a structure $\hat{G}(\mathcal{B}) = (\hat{G}(B), \Psi_{\hat{G}})$ (in $\underline{\underline{C}}^{\underline{\underline{E}}^{o}}$) and similarly for interpretations $\mathcal{B}^*$. Moreover, theorem 1 guarantees that every interpretation $\hat{G}(\mathcal{B})^*$ associated with $\hat{G}(\mathcal{B})$ is of the form $\hat{G}(\mathcal{B}^*)$ for a unique $\mathcal{B}^*$ associated with $\mathcal{B}$. This, in turn, yields a "uniform family of structures":

$$\mathcal{B}_X = G(X, \mathcal{B}) = (G(X, B), \Psi'_X) \qquad (X \in \mathrm{Ob}\ \underline{\underline{E}}),$$

whose associated "uniform families of interpretations" $\mathcal{B}^*_X$ $(X \in \mathrm{Ob}\ \underline{\underline{E}})$ are in a one-one correspondance with the $\hat{G}(\mathcal{B})^*$'s and ultimately with the $\mathcal{B}^*$'s, associated with $\mathcal{B}$.

I am sorry, but I need one more definition:

$\overset{X \in \underline{\underline{E}}}{\underset{\mathcal{B}_X}{\models}} H$ (read: <u>H is uniformly true in the</u> $\mathcal{B}_X$'s, $X \in \mathrm{Ob}\ \underline{\underline{E}}$) iff there is a $(\mathcal{B}^*_X)_{X \in \mathrm{Ob}\ \underline{\underline{E}}}$, uniform, such that $\underset{\mathcal{B}^*_X}{\models} H$, for all $X \in \mathrm{Ob}\ \underline{\underline{E}}$.

We are now in the position to state the main theorem:

THEOREM 3. <u>An SH is true in</u> $\mathcal{B}$ <u>iff it is uniformly true in the</u> $\mathcal{B}_X$'s:

$$\underset{\mathcal{B}}{\models} H \quad \underline{\text{iff}} \quad \overset{X \in \underline{\underline{E}}}{\underset{\mathcal{B}_X}{\models}} H.$$

In order to prove this theorem I will need several lemmas, which, but for one, I will state without pboof. What will be proven will give a sample of the (rather elementary) techniques involved. The complete proofs can be found in [6].

LEMMA 1. - Let $\mathcal{C}^* = (C, \Psi^*)$ be any interpretation (in $\underline{C}$), let H_0 be an atomic SH of rank $\underline{n}$ and let $g: Y \longrightarrow C^n$. Then

$$\models_{\mathcal{C}^*} H_0 [g] \quad \text{iff for every} \quad x: U \longrightarrow Y, \quad \models_{\mathcal{C}^*} H_0 [gx].$$

In the following lemmas, let $\mathcal{B}^* = (B, \Psi^*)$ be an interpretation in $\underline{D}$ and let H_0 be an atomic SH of rank $\underline{n}$.

LEMMA 2. - For $g: Y \longrightarrow B^n$, get $\hat{g}: U \longrightarrow G(KY, B)^n$ via the following natural transformations:

$$\frac{\dfrac{\dfrac{Y \longrightarrow B^n}{KY \longrightarrow KB^n}}{U \longrightarrow G(KY, B^n)}}{U \longrightarrow G(KY, B)^n.}$$

Then $\models_{\mathcal{B}^*} H_0 [g]$ iff $\models_{\mathcal{B}^*_{KY}} H_0 [\hat{g}]$.

LEMMA 3. - Let $X \in \text{Ob}\ \underline{E}$. For $g: Z \longrightarrow G(X, B)^n$ and $x: U \longrightarrow Z$, since $gx: U \longrightarrow G(X, B)^n$, get $g_x: X \longrightarrow (KB)^n$ using the following natural bijections:

$$\frac{\dfrac{U \longrightarrow G(X, B)^n}{U \longrightarrow G(X, B^n)}}{X \longrightarrow K(B^n).}$$

Then $\models_{K(\mathcal{B}^*)} H_0 [g_x]$ (in $\underline{E}$) iff $\models_{\mathcal{B}^*_X} H_0 [gx]$ (in $\underline{C}$).

LEMMA 4. - <u>For</u> $g: X \longrightarrow (KB)^n$ (in $\underline{\underline{E}}$), <u>get</u> $\bar{g}: FX \longrightarrow B^n$ <u>through isomorphism</u> $(KB)^n \approx K(B^n)$ <u>and adjunction</u> $F \dashv K$. <u>Then</u>

$$\models_{\mathcal{B}^*} H_0[\bar{g}] \qquad \underline{\text{iff}} \qquad \models_{K(\mathcal{B}^*)} H_0[g].$$

REMARK. - <u>Lemma 4 is the only place where adjunction</u> $F \dashv K$ <u>is used</u>.

COROLLARY 3. - <u>If H is true in every</u> $\mathcal{B}^*_X$, <u>then H is true in</u> $\mathcal{B}^*$.

PROOF. - Take $X = KY$ and apply lemma 2.

COROLLARY 4. - <u>Let</u> $g: Z \longrightarrow G(X, B)^n$. <u>Using the notations of lemmas 3 and 4, the following holds:</u>

$$\models_{\mathcal{B}^*_X} H_0[g] \quad \underline{\text{iff}}, \ \underline{\text{for every}} \ x: U \longrightarrow Z, \ \models_{\mathcal{B}^*} H_0[\bar{g}_x].$$

COROLLARY 5. - <u>If H is true in</u> $\mathcal{B}^*$ <u>and</u> $X \in \mathrm{Ob}\,\underline{\underline{E}}$, <u>then H is true in</u> $\mathcal{B}^*_X$.

PROOF. Only for H elementary, say $H_1 \wedge \ldots \wedge H_r \longrightarrow H_0$. Let $g: Z \longrightarrow G(X, B)^n$ be such that $\models_{\mathcal{B}^*_X} H_1 \wedge \ldots \wedge H_r[g]$. Then, for every $x: U \longrightarrow Z$, $\models_{\mathcal{B}^*} H_i[\bar{g}_x]$ by corollary 4. H being true in $\mathcal{B}^*$, it follows that $\models_{\mathcal{B}^*} H_0[\bar{g}_x]$ and hence $\models_{\mathcal{B}^*_X} H_0[g]$, again by corollary 4.

We can now prove theorem 3, thus ending the present lecture.

Assume $\models_{\mathcal{B}} H$ and let $\mathcal{B}^*$ be an interpretation associated with $\mathcal{B}$ such that $\models_{\mathcal{B}^*} H$. By the remark following corollary 2, $\mathcal{B}^*$ yields a uniform family $\mathcal{B}^*_X$ and $\models_{\mathcal{B}^*_X} H$, by corollary 5. Hence $\overset{X \in \xi}{\models_{\mathcal{B}_X}} H$.

Conversely, assume $\overset{X \in \xi}{\models_{\mathcal{B}_X}} H$ and let $\mathcal{B}^*_X$ be a uniform family such that $\models_{\mathcal{B}^*_X} H$, all X's. By the above remarks, there is a $\mathcal{B}^*$ associated with $\mathcal{B}$ such that $\mathcal{B}^*_X = G(X, \mathcal{B}^*)$. Now corollary 3 yields the desired result.

REFERENCES

[1] . COSTE, Logique du 1er ordre dans lès topos élémentaires. (Mimeographed notes).

[2] C. MARCHINI, Funtori che conservano e riflettono le SH, Atti Sem. Mat. e Fis. Univ. Modena 22 (1973).

[3] C. MARCHINI, Alcune questioni di semantica categoriale (to appear).

[4] M. SERVI, Una questione di teoria dei modelli nelle categorie con prodotti finiti, Matematiche (Catania) 26.

[5] " " , Su alcuni funtori che conservano le SH, Riv. Mat. Univ. Parma (3) 3 (1974).

[6] " " , A generalization of the exponential functor in connection with the SH-formulas (to appear).

CENTRO INTERNAZIONALE MATEMATICO ESTIVO

(C.I.M.E.)

TOPOLOGICAL MODEL THEORY

J. A. MAKOWSKY

Corso tenuto a Bressanone dal 20 al 28 giugno 1975

TOPOLOGICAL MODEL THEORY

J.A.Makowsky

CONTENT

TOPOLOGICAL MODEL THEORY : A SURVEY

J.A.Makowsky*

Istituto Matematico Firenze

II.Mathematisches Institut der

Freien Universität Berlin

0. INTRODUCTION

Topological model theory is getting "en vogue". It stems from the fact that model theory has been very successfull for algebraic structures, clarifying algebraic concepts (algebraic closure, Lefshetz principles etc) and the concept of infinitesimals (non-standard analysis) and recently is even invading "hard" algebra (Whitehead's conjecture).
It was always considered unsatisfactory that topological spaces could not be treated within the model theory of first order languages. It was usually argued that topology is basically a second order concept and even in ROBINSON's book "Non-standard analysis" [30] non-standard topologies are non-standard models of the full structure over the topological space.
Until recently, say 197o, few attempts were made to attack the problem. It was again ROBINSON who now proposed it as a problem [31]:

* Partially supported by the Italian C.N.R (G.N.S.A.G.)

"The task I which to specify here is of general nature: It is to develop TOPOLOGICAL MODEL THEORY in the direction exemplified above ([31]) or in any other direction " His approach was to consider structures $\mathfrak{A}$ equiped with a topology where all the functions of $\mathfrak{A}$ are continuous and all the relations either open or closed. PETRESCU [27] proved compactness and Löwenheim-Skolem theorems for this approach and ROBINSON, in one of his last papers [32],studied continuity of definable predicates and Skolem functions.

In this paper I try to give a survey of what has been done up to now in "any other direction", and I hope to show that there is a very good notion of first order topology. This paper is an updated and extended version of my talk given at the CIME 1975 .

The organization of the paper is more or less chronological. Very briefly the approach by HENSON,JOCKUSCH,RUBEL and TAKEUTI [10] will be discussed. Their paper "First order topology" is full of topologically interesting ideas, but it does not seem that their approach could fit into a program for model theoretic investigations. The more successfull approach starts with SGRO's thesis [33], where a generalized quantifier (the open set quantifier) was introduced. There techniques developed by KEISLER to study the quantifier "there exist uncountably many" [12 were exploited. These techniques were also successfully applied by the author and ZIEGLER [23] to construct the logic L(I) which has a new logical operation inspired by CHANG's paper "Modal model theory"[3] to deal with the interior of a definable set. L(I) behaves essentially exactly as first order logic, but it is too weak to express many interesting topological concepts. What makes these and related approaches successfull was captured first by GARAVAGLIA [7] who discovered the general principle behind all this: The notion of topological invariance for two sorted languages. A syntactical characterization of invariant sentences, similar to LEVY's approach to set theory via absoluteness, was first given by ZIEGLER [40]. Virtually all local concepts in topology are definable with invariant sentences but their logic has still most of the features of firts order logic. More surprising even, it can be characterized as the strongest topological logic satisfying these properties exactly as in LINDSTRÖM's theorem. Finally it turns out that the same program works for a treatement of uniform spaces and a glance is given also to proximity spaces and closure spaces. The paper concludes with a brief discussion of decidability questions and a list of problems.

I said that topological model theory is getting "en vogue". Many of these ideas are in the air. My presentation is biased by my own work and the work I heard of personally. Priority questions occur whenever a new branch develops. But the main priority belongs to the late A.ROBINSON whoadvocated the undertaking of this project. Without his encouragement, directly and indirectly, few of us would have concentrated our efforts so much to contribute to the developement of topological model theory. The CIME summerschool on model theory 1975 was dedicated to the memory of A.ROBINSON who personally has contributed much to the developement of mathematical logic in Italy and it is therefore a special pleasure for me to contribute this paper in this volume. But this survey could not have been written without the many discussions I had with A.Macintyre and J.Sgro,with P.Mangani,A.Marcja and S.Tulipani and with M.Ziegler and W.Rautenberg.

1. THE LATTICE OF CLOSED SETS

The first serious attempt to study the model theory of topological spaces, or rather to introduce the notion of elementarily equivalent topological spaces, was suggested in a paper by HENSON, JOCKUSCH, RUBEL and TAKEUTI [10] entitled " First order topology".

Given a topological space X they consider the structure $\mathfrak{X} = \langle c(X), \subset \rangle$ where c(X) is the lattice of closed subsets of X and $\subset$ is set inclusion. Standard structures are structures which come from topological spaces, i.e. structures $\mathfrak{A} = \langle A', \subset \rangle$ such that there exist a top. space A with $c(A) \simeq A'$ up to lattice isomorphisms. A' therefore is always a complete lattice, a property which is obviously <u>not</u> first order. The question the four authors are mainly interested in is as follows: Given any space X, what other spaces Y are elementarily equivalent to X, i.e. for what spaces Y $c(X) \equiv c(Y)$ holds ? The following sample theorems are due to SWETT and FLEISSNER [38], [39] :

Let R be a topological space homeomorphic to the real line.

<u>THEOREM 1.1</u> If X is a topological space elementarily equivalent to R then

(i) X is first countable and σ-compact and its topology is induced by a dense, complete linear order without endpoints

(ii) X has precisely continuum many points and

(iii) if X is locally separable then X is homeomorphic to R.

Let S be a topological space homeomorphic to a Suslin line, i.e. a dense, linear, complete order with the order topology and such that every family of disjoint intervals is countable.

THEOREM 1.2 If S is a Suslin line then

(i) S is (∞,ω)-equivalent to R

(ii) it is provable in ZFC that there are T_1-spaces X which are not Suslin lines and X is elementarily equivalent to R.

Many topological properties of spaces are expressible within this framework, so connectedness, most separation properties, inductive dimension n, but topological compactness is not. Still there are compact spaces which are elementarily equivalent only to compact spaces. The main defect is that certain model theoretically important theorems fail in this context. For example the Skolem-Löwenheim theorem fails and its generalizations or the behaviour of the Hanfnumbers is governed in part by large cardinal axioms . Obviously the compactness theorem fails,too. From the point of view of decidability the situation is even worse: The set of sentences true in all T_1-spaces is at least as complicated as second order number theory. For more the reader should definitely consult [10], where he may find also an interesting problem list.

Although the research along these lines is very interesting and challenging it is clear that it does not give any reasonable first order language for topology. It rather resembles the investigations on the elementary theory of the lattice of degrees of unsolvability.
To find a "reasonable" first order language for topology is the subject of the rest of this paper.

2. THE OPEN SET QUANTIFIER

The first logic for topological structures which deserves the name "first order" was suggested by KEISLER and developed in SGRO's thesis [33]. It is derived from a way of looking at generalized quantifiers. Let L(Q) be the set of finite formulas formed as usually but with an additional clause:

(Q) If x is a variable and ϕ a formula so is $Qx\phi(x)$.

Models for L(Q) are structures of the form $\mathfrak{A} = < \mathfrak{A}_o, q >$ where $\mathfrak{A}_o$ is an ordinary L-structure (for first order logic) and q is any family of subsets of $|\mathfrak{A}_0|$. Such structures where extensively studied by KEISLER [12] and were there called <u>weak models</u>. Satisfiability for L(Q) is explained as usually with the additional clause

(*Q) $< \mathfrak{A}_o, q > \models Qx\,\phi(x,\bar{a})$ iff $\{ c \in |\mathfrak{A}_0| : \mathfrak{A} \models \phi(c,\bar{a})\} \in q$.

Various interpretations for L(Q) have been considered restricting the choice of q. If q consists of the uncountable sets of $|\mathfrak{A}_0|$ we get the quantifier "there exist uncountably many" which has been studied by many logicians. In SGRO's [34] and [35] q varies over topologies, i.e. families of open sets on $|\mathfrak{A}_0|$. Such structures are called <u>topological structures</u> . In contrast to ROBINSON's approach [32] no restrictions are made on the functions and relations of $\mathfrak{A}$.

A formula of L(Q) is <u>topologically valid</u> if it is true in all topological structures. SGRO gives the following axiomatization for the topologically valid formulas:

$Qx(x=x)$ & $Qx(x\neq x)$, $(Qx\,\phi$ & $Qx\,\psi) \to Qx(\phi \& \psi)$, $\forall y Qx\,\phi(x,y) \to Qx\exists y\,\phi(x,y)$

together with all the valid formulas and rules (and schemas) of classical predicate calculus. A formula is called topologically derivable if it can be derived by a finite proof in the above system. Consistency is defined as usually.

THEOREM 2.1 (SGRO) (i) A set T of formulas of L(Q) is topologically consistent iff it has a topological model.

(ii) The set of topologically valid formulas of L(Q) is recursively enumerable.

(iii) A set T of formulas of L(Q) has a topological model iff every finite subset of T has a topological model.

(iv) A set T of formulas of cardinality $\kappa \geq \omega$ which is consistent has a topological model of cardinality κ whose topology has a basis of cardinality κ.

(v) A set T of formulas of L(Q) is consistent with $\forall x Q y (x \neq y)$ iff T has a 0-dimensional normal topological model.

SGRO also gives an ultraproduct construction for topological models which satisfies an analogue of Łos'slemma.

Sofar this logic with the open set quantifier behaves very nicely in the sense that its model theory can be developed along the lines of classical model theory. But there are two defects. The expressive power of L(Q) does not satisfy any reasonable demands of a topologist and there is no interpolation or definability theorem for L(Q). Among the few topological concepts definable in L(Q) we have "T_1-separation" and "discrete space" if equality is part of the logic, but already "T_o-separation" or "Hausdorff" , "the interior of a definable set is not empty" or "open in the product topology","continuity of definable functions" are not definable in L(Q),cf. MAKOWSKY-MARCJA [20] and MAKOWSKY-ZIEGLER [23]. SGRO first found a counterexample to CRAIG's interpolation theorem for L(Q) which he extended together with the author to a counterexample to BETH's theorem using an idea of SHELAH (cf. MAKOWSKY-SHELAH [21]).

3. PRODUCT TOPOLOGIES

Let us now consider n-ary quantifier symbols for every finite n. We denote these symbols by Q^n and add a new formation rule to L(Q):

(Q^n) If $x_1, x_2, \ldots, x_n$ are distinct variables and ϕ is a formula so is $Q^n x_1 x_2 \ldots x_n \phi$. For brevity we write also $Q^n \bar{x}\, \phi$.

We denote by $L(Q^{<\omega})$ the set of formulas where we have one n-ary quantifier for every $n<\omega$ and have formed finite formulas with the usual rules for predicate calculus and with (Q^n) . The intended interpretation of these quantifiers is "$\phi(x_1, x_2, \ldots, x_n)$ is open in the product topology", so structures are the same as in the preceeding section, i.e. topological structures, and satisfiability is defined by the clause:

($*Q^n$) $\langle \mathfrak{A}_0, q\rangle \models Q^n \bar{x}\phi(\bar{x},\bar{a})$ iff $\{\bar{c} \in |\mathfrak{A}_0| : \mathfrak{A} \models \phi(\bar{c},\bar{a})\}$ is open in the product topology induced by q.

Similar approaches do exist for other interpretations than toplogies: MALITZ and MAGIDOR [16] considered a product quantifier for cardinality logics and MAKOWSKY and TULIPANI [22] for the more general monotone quantifiers. In fact if q is monotone (which is not the case for open sets) there is a canonical way to define satisfiability for $L(Q^{<\omega})$ by

$(*Q^n, \mathrm{Mon})$ $\langle \mathfrak{A}_0, q \rangle \models Q^n \bar{x}\, \phi(\bar{x}, \bar{a})$ iff there exists a $B \in q$ such that

$B^n \subset \{ \bar{c} \in |\mathfrak{A}_0|^n : \mathfrak{A} \models \phi(\bar{c}, \bar{a}) \}$.

For monotone quantifiersmuch of classical model theory can be adapted whereas the topological product quantifier seems less easy going. The use of the topological productquantifier is its greater expressive power: "Hausdorff" e.g. now is definable. SGRO [35] has axiomatized $L(Q^{<\omega})$ for the topological interpretation, but we wont give the axioms here since they are rather complicated. He also proved compactness and Löwenheim-Skolem theorems similarily as in the previous section. Again the interpolation and definability theorems fail. Applications exist for the axiomatization of topological groups and vectorspaces and similar results hold for extensions of $L(Q^{<\omega})$ by infinitary conjunctions and disjunctions and by the cardinality quantifier "for uncountably many".

4. THE INTERIOR AND CHANG'S MODAL OPERATOR

In his paper "Modal model theory" [3] CHANG introduces a new type of a logical operation . He replaces the impersonal "it is necessary that ϕ" of classical modal logic by the personal " x finds it necessary that ϕ". A similar approach should work for a statement like "x is in the interior of the set defined by ϕ". So let us define the formal language L(I). L(I) is, like L(Q), an extension of classical predicate calculus where we have a new logical symbol I. The usual formation rules for finite formulas are extended by

(I) If x is a variable and ϕ a formula of L(I) so is $Ix\phi$.

x occurs free in $Ix\phi$, even if x does not occur in ϕ.

Weak models for L(I) are defined similarily as for L(Q), but with a local touch: the family q is replaced by a system of neighborhoods N_a for every element of the structure. More precisely structures are of the form $\mathfrak{A} = \langle \mathfrak{A}_0, N_a, a \in A \rangle$ where $\mathfrak{A}_o$ is a classical L-structure with universe A and N_a is a family of subsets of A. The additional clause in the definition of satisfiability now reads

(*I) $\mathfrak{A} \models Ix\phi(x,\bar{a})|_c$ iff $\{ b \in A : \mathfrak{A} \models \phi(b,\bar{a}) \} \in N_c$.

In [3] CHANG studies the model theory for weak models for L(I) and also for a slight generalization. He suggests several interpretations for L(I) which all come from modal or deontic logic. A semantic for an interpretation which is related to the modal system S4 was proposed by MAKOWSKY and MARCJA in [19] where a completeness theorem is proved. We call a weak model for L(I) a topological model if the N_a's are "real" neighborhoods,i.e. if $X \subset A$ and $X \in N_a$ so $a \in X$ and N_a is a filter in the power set of A. The model theory for L(I) for topological structures was initiated by the author [17] and further developed in MAKOWSKY and ZIEGLER [23], where all the results of this section come from.

It is obvious that topological models for L(Q) can be transformed canonically into topological models for L(I) and vice versa, therefore L(Q) is a sublogic of L(I) for the topological interpretation. L(I) is strictly stronger , but not too much : "T_o-space"is definable in L(I) but not in L(Q) and "the interior of a definable set", but "T_2-space" or "open in the product topology" are still not definable in L(I). In contrast to the usual quantifiers one should note that $IxIy\phi$ is not logically equivalent to $IyIx\phi$. A formula of $L(I)$ is topologically valid if it is true in all topological structures. The set of topologically valid sentences can be axiomatized by a Gentzen type system. If S and T are any finite sets of formulas we denote sequents by $S \to T$, where a sequent is valid iff $\sum_{s \in S} s \to \prod_{t \in T} t$ is valid. In addition to the usual rules (cf. LYNDON [15, pp 64 ff.]) we add two rules for I :

I-introduction on the left side $\dfrac{S,\ \phi(x) \to T}{S,\ Ix\phi(x) \to T}$ and

I-introduction on the right side

$$\frac{S,\ Ix\phi_1(x),\ Ix\phi_2(x),\ldots,\ Ix\phi_n(x) \to T,\ \phi(x)}{S,\ Ic\phi_1(c),\ Ic\phi_2(c),\ldots,\ Ic\phi_n(c) \to T,\ Ic\phi(c)}$$

for any term c provided x does not occur free anywhere else than indicated. Provability and consistency are defined as usually for Gentzen systems.

THEOREM 4.1 A set of sentences T in $L(I)$ is topologically consistent iff it has a topological model.

With similar methods as developed in SGRO [39] one proves the complete analogue to theorem 2.1 for $L(I)$, one gets an ultraproduct construction and most positive results which hold for $L(Q)$. The particular axiomatization has furthermore two definite advantages : One can prove an "omitting types theorem" via consistency properties and one can prove, by straightforeward inspection, the LYNDON interpolation theorem :

THEOREM 4.2 Let ϕ, ψ be formulas of $L(I)$ such that $\phi \to \psi$ is topologically valid. Then there exists a θ in $L(I)$ which contains only those extralogical symbols that occur both in ϕ and ψ and such that both $\phi \to \theta$ and $\theta \to \psi$ are topologically valid. Furthermore a relationsymbol occurs positively (negatively) in θ only if it occurs positively (negatively) in both ϕ and ψ. (Note that = is a logical symbol).

From this the usual BETH definability theorem and the following preservation theorem are derived.

THEOREM 4.3 A formula of L(I) is preserved under continuous open homomorphisms iff it is topologically equivalent to a positive formula. (In the definition of positive formula no restriction on the I-operator are made).

For the proof of the next theorem a method similar to the method used in NEBRES [26] is applied.

THEOREM 4.4 Let $\mathfrak{A}$ and $\mathfrak{B}$ be two structures such that $\mathfrak{A}_o$ is a substructure of $\mathfrak{B}_o$ and A is open in $\mathfrak{B}$. Then the following are equivalent

(i) If a formula ϕ of L(I) holds in $\mathfrak{B}$ then it holds in $\mathfrak{A}$.

(ii) ϕ is topologically equivalent to a universal formula of L(I) (where in the definition of universal formula no condition on the I-operator is made).

Another advantage of L(I) over L(Q) with respect to topological structures is a "back and forth" characterization of elementary equivalence. In the game theoretic description the game is a localized version of a game for the cardinality quantifiers and uses the observation that this game gives a definition of partial isomorphisms which fits with the syntax if the quantifier is monotone (cf . MAKOWSKY-TULIPANI [22]). The "back and forth" technique can be used to get FEFERMAN-VAUGHT type theorems such as"elementary equivalence is preserved under sums and products

in the topological sense." With the above mentioned "omitting types theorem" and the "back and forth" technique one gets easily a characterization of $\aleph_0$-categorical theories, exactly as in the ERNS-theorem (ENGELER, RYLL-NARDZWESKI, SVENONIUS). Here a L(I)-theory T is $\aleph_0$-categorical if it is complete and two countable topological models with the coarsest compatible topology are homeomorphic.
This last result , and most classical results of model theory as MORLEY's theorem , can be obtained by a different method. L(I) can be reduced to classical predicate calculus in the sense that for every topological model $\mathfrak{A}$ there is an expansion $\mathfrak{A}^*$ of $\mathfrak{A}_0$ by definable relations and constants for every $a \in A$ and a translation of formulas ϕ of L(I) into formulas ϕ^* of predicate calculus in the expanded language such that

$\mathfrak{A} \models \phi$ iff $\mathfrak{A}^* \models \phi^*$.

Obviously L(I) can be extended further to $L(I^{<\omega})$ as L(Q) in section 3 and in fact similar theorems hold. But for $L(I^{<\omega})$ it is open wheter the interpolation and definability theorems hold.
The reason why so many topological logics do exist lies in the fact that there is a common extension L^t which can be characterized similarily as predicate calculus in LINDSTRÖM's theorem [14].

5. A MAXIMAL LOGIC FOR TOPOLOGICAL STRUCTURES

All the topological concepts which were definable in one of the topological logics treated so far have one thing in common: Their topological definition does not depend generally on open sets but only on the existence of some set in a basis for the topology. A way to exploit this was first discovered by GARAVAGLIA [7, 8, 9]: Let L_2 be the two-sorted first order language appropriate for structures $< \mathfrak{A}, \alpha, \in >$ where $\mathfrak{A}$ is an L-structure and α is a set of subsets of $A = |\mathfrak{A}|$. We call such a structure <u>topological</u> if α is a topology. We call such a structure <u>basic</u> if α is closed under finite intersection. If β is any family of subsets of A we define $\tilde{\beta}$ to be $\{ \cup s : s \subset \beta \}$. We call a formula ϕ <u>B-invariant</u> if for every L_2-structure $< \mathfrak{A}, \beta, \in > = \underline{\mathfrak{A}}$ ϕ holds in $\underline{\mathfrak{A}}$ iff it holds in $< \mathfrak{A}, \tilde{\beta}, \in >$. Note that $\tilde{\beta}$ is a topology on $\underline{\mathfrak{A}}$ iff $\underline{\mathfrak{A}}$ satisfies the sentence <u>Top</u> which is the formula

$\forall x \exists X (x \in X) \& \forall x \forall X (x \in X \Rightarrow \forall Y (x \in Y \Rightarrow \exists Z (x \in Z \& \forall y (y \in Z \Rightarrow y \in X \& y \in Y))))$. GARAVAGLIA defined his topological logic to be the set of B-invariant sentences which are consistent with <u>Top</u>.

We now define the logic L^t which is independently due to GARAVAGLIA and ZIEGLER [40]. We follow here ZIEGLER's approach. Whenever a theorem is attributed to both GARAVAGLIA and ZIEGLER, this means that GARAVAGLIA has proved the theorem first or independently for the B-invariant sentences. L^t is the smallest subset of L_2 above where the existential set quantifier occurs only in the following form $\exists X (t \in X \& \phi(X))$ where t is a term in L_2 and X does not occur positively in ϕ. (X occurs positively in ϕ if a free occurence of X in ϕ is inside the scope of an even number of negation symbols. If X has no free occurence in ϕ then X does not occur positively.) Note that $\forall X$ is defined as $\sim \exists X \sim$.

<u>LEMMA 5.1</u> (GARAVAGLIA, ZIEGLER) All the formulas of L^t are B-invariant.

For the rest of this section we consider only topological models.

<u>THEOREM 5.2</u> (GARAVAGLIA, ZIEGLER)

(i) A theory T in L^t has a topological model iff it is consistent with <u>Top</u>.

(ii) The set of L^t-sentences true in all topological models is recursively enumerable.

(iii) L^t satisfies the compactness theorem for sets of formulas of arbitrary cardinality.

(iv) L^t satisfies the downward Löwenheim-Skolem theorem: A set T of sentences in L^t with $|T| = \kappa$ which has a model has a model of cardinality $\leqq \kappa$ with a basis of cardinality $\leqq \kappa$.

There is a natural extension of the usual EHRENFEUCHT-FRAISSE characterization of elementary equivalence and of partial iso- (here homeo-) morphisms. In particular, in contrast to the similar characterization for L(I) in section 4, the L^t-partial isomorphisms do converge to homeomorphisms in the sense of FLUM [6] and therefore we have

<u>THEOREM 5.3</u> (ZIEGLER) Let L^* be any logic for topological structures which satisfies

(i) L^* does not distinguish homeomorphic structures

(ii) L^* satisfies the downward Löwenheim-Skolem theorem for countable sets of formulas

(iii) L^* is countably compact.

Then L^* is a sublogic of L^t.

The definition of "any logic..." is completely analogue to the one given in KEISLER's paper in this volume [13] adapted for many-sorted structures. As a corollary one gets immediately:

COROLLARY 5.4 (ZIEGLER) A sentence ϕ of L_2 is B-invariant iff

$$\text{Top} \vdash \phi \leftrightarrow \psi \quad \text{for some sentence } \psi \text{ of } L^t .$$

There are several points which make us believe that L^t is a natural logic for topological structures :

- All the logics treated in sections 2-4 are sublogics of L^t. Furthermore the concept of a continuous function or an open relation is expressible in L^t. Therefore L^t also fits ROBINSON's program for topological model theory.
- L^t is maximal with respect to compactness and the downward Löwenheim-Skolem property. Now these two properties are generally considered as a good substitute for a formal definition of "first order". So this suggests that no extension of L^t is"reasonably first order".
- A good deal of classical model theory can be developed also for L^t.

The rest of this section is devoted to examples of such model theoretic results.

Definability theory. Using the construction behind the proof of Lindström's theorem or its analogue theorem 5.3 one can prove CRAIG's interpolation theorem and from this one can derive BETH's definability theorem and related results. These results also follow using the ultraproduct construction below. But a proof theoretic argument was not yet found.

Ultraproducts . An ultraproduct construction for the B-invariant sentences was first given by GARAVAGLIA. He and ZIEGLER proved independently that elementarily equivalent (in L^t) structures have homeomorphic ultrapowers. As an application (via a result due to BANKSTON [1]) one gets that the theory of perfect T_3-spaces is complete and another proof of corollary 5.4.

Categoricity. We call a theory T κ-categorical (in L^t) if two models of T of cardinality κ are homeomorphic.

THEOREM 5.5 (ZIEGLER) Any countable perfect T_3-space is homeomorphic to the rational numbers with the standard topology, hence the theory of perfect T_3-spaces is $\aleph_o$-categorical.

For L^t it is still open wheter a" good omitting types theorem" holds to prove interesting categoricity theorems (as the ERNS- or MORLEY's theorem), but as long as there are no further natural examples of categorical topological theories, there is no interest in such theorems.

Preservation theorems. Using the previously mentioned back and forth technique one easily gets the classical results that sums and products preserve elementary equivalence in L^t. ZIEGLER also obtained syntactical characterizations of the sentences preserved under topological homomorphisms and substructures etc.

Decidability questions will be discussed in section 7, but let us mention that the set of B-invariant sentences in L_2 is not r.e.

Let us finish this section with one more general remark:

As one would expect in a first order logic, many important topological concepts are not definable in L^t, among those compactness, connectedness, separation above T_3 etc. But similar remarks apply to group theory and classical model theory. The question arises what use topological model theory might have for topology ? In my personal opinion model theory might lead topologists to ask a question so far neglected in topology: How many different (compatible) topologies of one or the other kind one may find on a given structure. As far as I could find out it is open wheter every infinite group admits a T_2-non-discrete topology ?

PODEWSKI [28, 29] and HESSE [11] investigated this problem for fields and almost abelian groups and PODEWSKI showed

THEOREM 5.6 (PODEWSKI) If a group (field) of cardinality κ admits a compatible T_2-non-discrete topology then it admits $2^{2^{\kappa}}$-many different topologies.

As an application of the compactness theorem of L^t one gets

THEOREM 5.7 A group G admits a T_2-non-discrete compatible topology iff every finitely generated subgroup of G does.

It may very well be that an adaptation of SHELAH's theory of stable diagrams (cf. his forthcoming book [36]) to L^t may help to solve problems in this direction. It might even be a source for almost unlimited inspiration in this direction.

6. UNIFORMITIES, PROXIMITIES AND CLOSURE SPACES

In the topological literature many other types of spaces exist which are in some way related to topological spaces. On one side there are weaker concepts, among which the closure spaces are the most interesting. On the other side there are stronger concepts which come mostly from analysis. Among these the uniform spaces and the proximity spaces are the best known. What all these spaces have in common is that they use different basic concepts which canonically induce topologies. In this section we shall give some indications how the previous approaches apply for the above spaces. Topological reference books are e.g CECH [2] and ENGELKING [5].

Uniform spaces. The basic notions for uniform spaces are uniformities and entourages. They play analogue roles as topology and neighborhood. The analogue to SGRO's approach in section 2 would be to consider a binary quantifiersymbol Uxyϕ which reads "the set defined by ϕ is an entourage". Structures then were pairs $< \mathfrak{A}, u>$ where $\mathfrak{A}$ is an ordinary structure and u is a uniformity. As one easily observes most of the results of section 2 carry over but the author did not check the details. Anyhow this approach seems rather pointless: its expressive power is very weak (as for L(Q)) and, that it works, just repeats an already made experience. More interesting, it seems, is the approach sketched in section 5. Structures now are triples of the form $< \mathfrak{A}, u, \varepsilon >$ where $\mathfrak{A}$ is a classical structure, u is a uniformity on $\mathfrak{A}$ and ε is a ternary relation on $A^2 \times u$ with $\varepsilon(a_1,a_2,v)$ iff the pair (a_1,a_2) is in v. If u' is an arbitrary family of subsets of A^2 we denote by $\tilde{u}'$ the set $\{ \bigcup w : w \subset u' \}$. With this a notion of B-invariance for L_2-formulas can be defined exactly as in section 5. L^u is the sublogic of L_2 (appropriate to the above structures) where existential quantification of members of u occur only in the form $\exists v \phi(v)$ where v does not occur positively in ϕ. With these definitions all the theorems of section 5 can be proved for L^u. A detailed exposition of this program may be found in STROBEL's thesis (Diplomarbeit) [37].

Proximity spaces. The basic notion of proximity spaces is nearness. A typical example is the following relation between sets of a topological T_4-space T: If A,B are subsets of T we say that they are near iff $cl(A) \cap cl(B) \neq \emptyset$. For the definition of an abstract proximity we refer to ENGELKING [5] or NAIMPALLY-WARRACK [25]. One way to develop a first order logic for proximity spaces is the following: Weak structures are pairs $<\mathfrak{A},\P>$ where $\mathfrak{A}$ is an ordinary first order structure and $\P$ is a subfamily of $\mathbb{P}(A)^2$. The logic L(P) is obtained from predicate calculus by adjunction of a binary quantifier symbol P and the formation rule

(P) If ϕ and ψ are formulas of L(P) and x,y are variables such that x (y) does not occur in ψ (ϕ) then $Pxy(\phi:\psi)$ is a formula of L(P).

Satisfiability is defined by the clause

(*P) $<\mathfrak{A},\P> \models Pxy(\phi:\psi)$ iff $(\phi(\mathfrak{A}),\psi(\mathfrak{A}))\in\P$.

A weak structure is called an LE-structure if $\P$ is a LEADER-proximity, a LO-structure if $\P$ is a LODATO-proximity and a P-structure if $\P$ is a proximity. (cf. NAIMPALLY-WARRACK [25]). L(P) with these three interpretations is studied in MAKOWSKY [18]. The logic L(P) is closely related to L(I), in fact in a certain sense they are equivalent. Consider the following translations of L(P)-formulas into L(I)-formulas:

$\ell_1(Pxy(\phi:\psi)) := \exists z(\phi|^x_z \& \sim Iz\sim \psi|^y_z)$ and

$\ell_2(Pxy(\phi:\psi) := \exists z(\sim Iz\sim\phi|^x_z \& \sim Iz\sim\psi|^y_z)$ and the translation from L(I)-formulas into L(P)-formulas $\ell(Iz\ \phi) := Pxy(z=x:\phi|^z_y)$.

THEOREM 6.1 (MAKOWSKY) If ϕ is a sentence in L(P) then

(i) ϕ has an LE-model iff $\ell_1(\phi)$ has a topological model

(ii) ϕ has an LO-model iff $\ell_2(\phi)$ is consistent in L(I) with R_o-separation $\forall xy(Ix(x\neq y)\Longleftrightarrow Iy(x\neq y))$.

(iii) ϕ has a P-model iff $\ell_2(\phi)$ is consistent in L(I) with T_1-separation $\forall xy(Ix(x\neq y)\Longleftrightarrow(x\neq y))$.

If ϕ is a sentence in L(I) then

(iv) ϕ has a topological model iff $\ell(\phi)$ has an LE-model.

As a consequence all the results of section 4 carry over to L(P) ,in particular the interpolation theorem.

Closure spaces. In his book [2] CECH observed that much of the theory of topological spaces may carried over to closure spaces, i.e. spaces with a closure operation cl satisfying the axioms $cl\emptyset = \emptyset$, $X \subset clX$, $cl(X \cup Y) = clX \cup clY$. It is easy to check that the approach via L(I) of section 4 is also good for closure spaces. Most results of section 4 carry over although no Gentzen system has been developed up to now. The only interest closure spaces have in this context is their role in a conjecture of JASKOWSKI which will be treated more closely in the next section.

7. DECIDABILITY

In this section some decidability problems for topological spaces will be discussed. Some of them are by now classical, but new proofs can be given with the use of model theory. We shall restrict ourselfs to decidability of topological spaces omitting important work on topological groups and fields. In particular SGRO is working on a project on topological groups where he already got interesting results. For an elimination technique of the I-operation and its application to topological fields the reader may consult MAKOWSKY-MARCJA [20].

Consider now the following language for topology:

We have only setvariables X,Y,... and the operations $\cap,\cup,\sim,cl$ (intersection,union,complement and closure) and the constant $\emptyset$. With this we may build finitary terms.Denote by T(cl) the set of so built terms. The equational calculus over T(cl) consists of the equations $t = t'$ for terms from

T(cl).Denote by E(cl) the set of equations true in all topological spaces and by C(cl) the set of equations true in all closure spaces with the obvious interpretation.

THEOREM 7.1 (i) (MCKINSEY-JASKOWSKI) E(cl) is decidable.

(ii) (MAKOWSKY) C(cl) is decidable.

If we replace the closure cl by a unary operation ' (derivative) we can consider the corresponding sets T(der),E(der) and C(der). In a footnote of MCKINSEY [24] it is claimed that JASKOWSKI had proved the decidability of E(der) but no proof was ever published. Let us therefore call this last statement JASKOWSKI's conjecture. Denote by E_1(der) the set of equations over T(der) true in all T_1-spaces.

THEOREM 7.2 (i) (ZIEGLER) E_1(der) is decidable.

(ii) (MAKOWSKY) C(der) is decidable.

The point of interest here is that both theorem 7.1 and 7.2 can be proved via the model theory of L(I) observing that the sets of equations above are subsets of the monadic L(I)-logic without equality (for cl) or with equality (for der). For 7.1 and 7.2(ii) one proves the stronger result that if an equation has a model it has a finite model where the cardinalitydepends only on the lenghth of the term occuring in the equation. (cf. MAKOWSKY-ZIEGLER [23]). 7.2(i) depends on the following

THEOREM 7.3 (ZIEGLER) The theory of T_1-spaces is

(i) decidable in L(Q)

(ii) decidable in L(I)

(iii) undecidable in L^t

whereas the theory of all topological spaces is

(iv) undecidable in L(Q),L(I) and L^t.

For L(Q) the result was also known to SGRO.

Using theorem 5.5 and the decidability of the monadic weak second order theory of two successor functions (DONER [4]) one gets

THEOREM 7.4 (ZIEGLER) The theory of T_3-spaces with a finite number of unary predicates is decidable in L^t whereas the theory of T_2- spaces is undecidable.

8. SOME OPEN PROBLEMS

In this section some research problems are presented. Again their choice is rather personal and biased by the authors own research. Most problems are not original :they are in the air or more precisely in the background of the many discussions I had with all the mathematicians mentioned in the introduction.

For problems concerning the lattice of closed sets the reader is refered to the excellent problem list in [10]. For L(Q) no particular problems are listed since most problems for L(I) below have their analogues for L(Q). Concerning the product quantifier the main problem is the absence of a general theory (or even orientation). Therefore we state

PROBLEM 1 Develop a general theory of product quantifiers. In particular, what properties of the unary quantifier carry over to the product quantifier for a suitable definition of the product (in the monotone case $(*Q^n,\mathrm{Mon})$) . Is it true that if the unary quantifier is axiomatizable for a given interpretation so are the product quantifiers for the corresponding product interpretation? What other model theoretic properties are preserved ?

Concerning L(I) the problems are more precise. In section 7 a connection between L(I) and the equational calculus of the closure and derivative operations was sketched.

PROBLEM 2 Prove or disprove JASKOWSKI's conjecture eventually using the model theory of L(I).

In section 4 two preservation theorems (4.3 and 4.4) were presented.

PROBLEM 3 Characterize the formulas of L(I) which are preserved under inductive limits (in the category of topological spaces). What are the formulas preserved under reduced products ?

L(I) has virtually all the properties of predicate calculus. It also has the following (technically) important property :

(DEF) Given a topological structure $<\mathfrak{A},q>$ let d be the family of the L(I)-definable sets using parameters from $\mathfrak{A}$. Then $<\mathfrak{A},q>$ and $<\mathfrak{A},\tilde{d}>$ are elementarily equivalent in L(I) (with parameters).

ZIEGLER now conjectures

CONJECTURE 4 Given any logic L^* for topological structures which satisfies the hypothesis of theorem 5.3 and (DEF). Then L^* is a sublogic of L(I).

More problems on L(I) may be found in MAKOWSKY-ZIEGLER [23].

ZIEGLER showed that the T_3-spaces with one equivalence relation is undecidable in L^t.

PROBLEM 5 Is the theory of T_3-spaces with one equivalence relation with closed equivalence classes decidable in L^t ?

The T_2-spaces are undecidable in L^t.

PROBLEM 6 Are the uniform Hausdorff-spaces decidable in L^u ?

L(I) has a nice proof theory as shown in section 4.

PROBLEM 7 Develop a "nice" proof theory for L^t or L^u.

PROBLEM 8 What kind of usefull omitting types theorem do hold for L^t or L^u ? What kind of categoricity theorems are true in L^t or L^u ?

Finally there two more very general problems :

PROBLEM 9 Extend the framework of abstract model theory (cf. KEISLER's paper in this volume [13]) such that topological model theory and even other model theories (KRIPKE models e.g.) fit nicely into it .

PROBLEM 1o What use for topology does topological model theory have ?

Added in January 1976:

From McKEE (Zeitschrift f.math.Logik und Grundlagen der Mathematik, vol.21 (1975) pp 4o5-4o8 , Infinitary logic and topological homeomorphisms) I can guess that most of the results described in section 5 were known to T. A. McKEE already in August 1974. I am glad that I could - better late than never - give credit for his work.

REFERENCIES

1. P.Bankston, Notices AMS, Abstract 75T-G37
2. E.Cech, Topological spaces, Prag 1966
3. C.C.Chang, Modal model theory, Proceedings of the Cambridge summer school in mathematical logic, Springer Lecture Notes in mathematics vol. 337 pp. 599-617
4. J.E.Doner, Notices AMS, Abstract 65T-468
5. R.Engelking, Outline of general topology, Warsaw 1968
6. J.Flum, First order logic and its extensions, Proceedings of the Logic conference,Kiel 1974, Springer Lecture Notes in Mathematics vol 499 (to appear)
7. S.Garavaglia, Notices AMS, Abstract 75T-E36
8. - , do , Abstract 75T-E61
9. - , do , Abstract 75T-E79
1o. C.W.Henson,C.G.Jockusch,L.A.Rubel and G.Takeuti, First order topology, preprint 1975
11. G.Hesse, Zur Topologisierbarkeit von verallgemeinerten abelschen Gruppen, Hannover 1974, preprint
12. J.Keisler, Logic with the quantifier "there exist uncountably many". Ann.math.logic vol.1.1 (1969) pp. 1-93
13. - , Constructions in model theory, this volume
14. P.Lindström, On extensions of elementary logic, Theoria vol. 35 (1969) pp. 1-11
15. R.Lyndon, Notes on logic, Princeton 1966

16. M.Magidor and J.Malitz, Compact extensions of L(Q), to appear in the Ann.math.logic

17. J.A.Makowsky, A logic for topological structures with an interior operator, Abstract presented at the SIECIL,Clermont-Ferrand 1975

18. - , A logic for proximity spaces, Abstract presented at the 1976 Spring meeting of the ASL.

19. - and A.Marcja, Completeness theorems for modal model theory with the Montague-Chang semantics I.,to appear in Zeitschr. f.math.Logik

2o. - , - , Problemi di decidibilità in logica topologica, to appear in Rend.d.sem.d.Univ.d.Padova

21. - and S.Shelah, The theorems of Beth and Craig in abstract model theory, to appear

22. - and S.Tulipani, Some model theory for monotone quantifiers, to appear

23. - and M.Ziegler, Topological model theory with an interior operator, to appear

24. J.C.C.McKinsey , A solution of the decision problem for the Lewis systems S.2 and S.4 with an application to topology, JSL vol.6 (1941) pp. 117-134

25. S.A.Naimpally and B.D.Warrack, Proximity spaces, Cambridge 197o

26. B.F.Nebres, Infinitary formulas preserved under unions of models, JSL vol.37.3 (1972) pp.449-465

27. J.Petrescu, O teorema de tip Löwenheim Skolem pentru modele topologice, Stud.Cerc.mat. vol.26 (1974) pp. 1237-1239

28. K.P.Podewski, The number of field topologies on countable fields, Proc.AMS vol. 39.1 (1973) pp. 33-38

29. - , Zur Topologisierbarkeit algebraischer Strukturen, Hannover 1974, preprint

3o. A.Robinson, Non-standard Analysis, Amsterdam 1974 (revised ed.)

- , Metamathematical problems,JSL vol. 38.3 (1973) pp. 5oo- 516

32. - , A note on topological model theory, FM vol.LXXXI.2 (1974) pp. 159-171

33. J.Sgro, Thesis, Madison 1974

34. - , Completeness theorem for topological models, to appear

35. - , Completeness theorems for continuous functions and product topologies, to appear

36. S.Shelah, Stability and the number of non-isomorphic models, Book, to appear

37. J.Strobel, Diplomarbeit, Berlin 1976

38. A.Swett, Notices AMS, Abstract 75T-E23

39. - and W.Fleissner, Notices AMS,Abstract 75T-E65

4o. M.Ziegler, A language for topological structures which satisfies a Lindström theorem, to appear in Bul.AMS

****

CENTRO INTERNAZIONALE MATEMATICO ESTIVO

(C.I.M.E.)

MODEL THEORY IN ALGEBRA WITH EMPHASIS ON GROUPS

G. SABBAGH

Corso tenuto a Bressanone dal 20 al 28 giugno 1975

Testo non pervenuto

Editoriale Grafica - Roma